T0257884

# Key Concepts of Proteomics

# Key Concepts of Proteomics

Edited by **Charles Malkoff**

New York

Published by Callisto Reference,
106 Park Avenue, Suite 200,
New York, NY 10016, USA
www.callistoreference.com

**Key Concepts of Proteomics**
Edited by Charles Malkoff

International Standard Book Number: 978-1-63239-441-5 (Hardback)

This book contains information obtained from authentic and highly regarded sources. Copyright for all individual chapters remain with the respective authors as indicated. A wide variety of references are listed. Permission and sources are indicated; for detailed attributions, please refer to the permissions page. Reasonable efforts have been made to publish reliable data and information, but the authors, editors and publisher cannot assume any responsibility for the validity of all materials or the consequences of their use.

The publisher's policy is to use permanent paper from mills that operate a sustainable forestry policy. Furthermore, the publisher ensures that the text paper and cover boards used have met acceptable environmental accreditation standards.

**Trademark Notice:** Registered trademark of products or corporate names are used only for explanation and identification without intent to infringe.

Printed in the United States of America.

# Contents

# Preface

The world is advancing at a fast pace like never before. Therefore, the need is to keep up with the latest developments. This book was an idea that came to fruition when the specialists in the area realized the need to coordinate together and document essential themes in the subject. That's when I was requested to be the editor. Editing this book has been an honour as it brings together diverse authors researching on different streams of the field. The book collates essential materials contributed by veterans in the area which can be utilized by students and researchers alike.

Proteomics is fundamentally defined as the extensive study of proteins, their structure as well as functions. It was considered to be an obvious next step in research once the field of genomics had deposited vital magnitude of information. However, taking an instantly verbatim move towards cataloging all proteins in all tissues of dissimilar organisms is not practically easy. Experts may need to focus on the aspects of proteomics that are necessary to the functional outcome of the cells. In this book, experts have presented the historical reviews of proteomics, as well as added viewpoints and fresh improvements in gel and non-gel-based protein partition and recognition by mass spectrometry. Extensive accounts have unfolded researches on sub-proteomes such as phosphoproteomes or glycoproteomes which are straightforwardly linked to functional outcomes of the cells. Structural proteomics connected to pharmaceutics growth is also an aspect noteworthy of consideration. Bioinformatics instruments that can extract proteomics information are also examined. This book is an ideal reference for students and even experts.

Each chapter is a sole-standing publication that reflects each author's interpretation. Thus, the book displays a multi-facetted picture of our current understanding of application, resources and aspects of the field. I would like to thank the contributors of this book and my family for their endless support.

**Editor**

# Part 1

# Proteomics – Historical Review

# Evolution of Proteomic Methods for Analysis of Complex Biological Samples – Implications for Personalized Medicine

Amanda Nouwens and Stephen Mahler
*The University of Queensland*
*Australia*

## 1. Introduction

We are on the threshold of a paradigm shift for proteomics, moving from largely a qualitative discipline, to now being capable of quantification of a protein within a complex sample at great sensitivity. The potential application of such advanced proteomic technology is enormous as we will be able to detect and quantify low levels of expressed proteins in complex samples, and so move comparative proteomics to a new level.

The evolving practice of personalized medicine will be dependent on devising new techniques and methodologies that will allow the detection and quantification of proteins that are implicated in contributing to the diseased state. There is perhaps somewhere over 5000 genes that are linked to disease states and complex networks of interactions of these expressed genes ultimately lead to these disease states. The myriad of single nucleotide polymorphisms (SNPs) contribute to such phenomena, as an individual's SNP profile play a major role in susceptibility to disease and in adverse reactions to drugs, for example. Coupled with mutations that occur throughout life, the complex "disease state" proteome will contain mutant proteins at low levels that need to be identified and quantified, so that therapeutic intervention based on rational scientific hypotheses can be investigated.

Plasma and serum contain an unknown number of proteins with amounts ranging from pg-g/L levels (i.e. very high dynamic range). As we know one of the major problems faced by proteomic studies of plasma or serum, or indeed any complex protein sample, is that a relatively small number of abundant proteins accounts for the great majority of protein content of the sample. The upshot is that the proteins of interest, which may have regulatory function, are masked by these abundant proteins, and non-targeted methods of proteomic analysis bias at the top end of the abundance scale. The development of new methods for quantifying low abundance proteins has evolved rapidly, concomitant with the evolution of powerful mass spectrometers of increasing sensitivity. The use of antibodies for targeting peptides prior to mass spectrometry analysis is becoming prominent, as a means of partitioning low abundance peptides away from peptides in the bulk sample.

This review will provide a broad overview of the evolution of proteomic methods to analyse biological samples, including Differential In-Gel Electrophoresis (DIGE), Isotope-Coded Affinity Tag (ICAT), Isobaric tags for relative and absolute quantification (iTRAQ), Stable isotope labeling with amino acids in cell culture (SILAC), Unique ion signature Mass

Spectrometry, Selected reaction monitoring (SRM)- based targeted mass spectrometry and Stable Isotope Standards and Capture by Anti-Peptide Antibodies (SISCAPA). Examples of the application of these methods to the identification of proteins involved in a variety of disease states and their implication for personalized medicine will be provided.

## 2. Overview of personalized medicine

Personalized medicine is designed medicine based on the genotype, or more specifically the SNP profile of individuals. Personalized medicine facilitates the selection of treatments best matched to the individual and disease phenotype (Marko-Varga et al., 2007). What are the main factors which contribute to genotype diversity? SNP is the overriding factor, reflecting past mutation, and occurs wherever there is more than one nucleotide when comparing two sequences. It is the spread of SNPs within genomes which contribute to our individuality, with an estimated 93% of genes containing an SNP (Chakravati, 2001). The individual "fingerprint" of SNPs reflects differences in susceptibility to disease and our varying response to drugs. Pharmacogenetics is the study of how these differences in genotype are manifested in inter-individual variation in response to drugs. The convergence of traditional pharmacogenetics with the relatively new discipline of human genomics has resulted in the evolution of pharmacogenomics (Weinshilbaum et al., 2004). The Pharmacogenetics and Pharmacogenomics Knowledge Base (PharmGKB) is an entity associated with the cataloging genes involved in modulating the response to drugs. The Pharmacogenomics Research Network (PGRN) is a collaboration of scientists studying the effect of genes on responsiveness to a wide variety of medicines (Altman, 2007). The PGRN is linked to the PharmGKB integrative database containing genetic and clinical information on participants in studies (http:www.pharmgkb.org). Thus the integration and availability of data associated with information at the genomic and transcriptomic levels are well developed and a valuable resource for researchers involved in the development of personalized medicine.

The incorporation of proteomics in the further development of the concept of personalized medicine is a more recent phenomenon, and has given rise to the area of pharmacoproteomics, which in essence studies how the proteome changes in response to a drug, and is a logical extension of the pharmacogenetics and pharmacogenomics (Jain, 2004). While pharmacogenetics and pharmacogenomics provide information at the level of the genome and transcriptome, pharmacoproteomics yields information on function (i.e. translational level), although it should be pointed out that small nuclear (sn) RNAs have relatively recently been shown to contribute to regulation of cellular processes and thus have a functional role (Mercer et al., 2009; Mattick, 2009). To this end proteomics and proteomic profiling of individuals' serum and tissues are becoming increasingly important in patient diagnosis and assessment, and together with pharmacogenetics and pharmacogenomics, will provide a more complete picture of the status of an individual, particularly at the functional level.

Identification of disease states can be based on genomic analyses. For example, identification of mutations in breast cancer genes BRAC1 and BRAC2 can be used in the diagnosis of breast cancer (Miki et al., 1994; Wooster et al., 1995). However, at present DNA alone does not necessarily reflect the physiological state of functioning cells and thus analysis of gene products, both RNA and protein are required. RNA expression in comparison to protein, is easier to perform, but transcript levels in the cell do not always reflect protein levels, nor account for post-translational modifications on proteins or alternative splicing events.

## 2.1 Significance

There are a number ways in which proteomics may be utilized in personalized medicine and more broadly in drug discovery, research and development. In its simplest application, proteomics has contributed to the discovery of disease biomarkers, clinical entities that define and /or predict normal and pathogenic states (Krejsa et al., 2006). Furthermore, the clinical response to treatment can be monitored through proteome profiling of relevant biomarkers. The clinical use of a biomarker is contingent on whether it is a validated biomarker, which ultimately depends on its clinical reliability and utility. Combinations of validated markers biomarkers can indicate surrogate endpoints that can predict clinical outcomes. On a more global protein expression level, comparative proteomics can generate patterns of protein expression or expression profiles, which may be utilized to define a specific physiological state, or diseased state.

One area where proteomics can yield information not possible by other means is the identification and localization of proteins in various cellular compartments and extracellular space. The paradigm that proteins have fixed locations within cells has recently proven to be simplistic, and that proteins have diverse functions depending on their cellular location. The identification of a protein outside of its known functional zone in cellular preparations was once thought to be due to rupture of cells/cell organelles and leakage of the protein into other fractions. However it is now known that proteins translocate between intracellular and extracellular compartments (Butler et al., 2009). This has enormous implications in drug targeting as the presence of a target in multiple locations may complicate therapy. For example, chaperone proteins, including HSP 10, 70 and 90 have now been shown to exist in extracellular locations, where it was once thought that the chaperone proteins were exclusively located intracellularly, to aid protein folding and carry out chaperone function. Inhibitors of HSP90α are in clinical trials for treatment of cancer (Banerji, 2009), however inhibition of the extracellular wound-healing function of HSP90α21 could be an undesired adverse effect.

## 2.2 Personalized medicine for cancer

Due to the great diversity of cancer types, and individual variation within specific tumours, cancer perhaps shows the greatest potential for development of personalized therapy. Cancer accounts for about 13% of all deaths annually world wide, and is a major cause of morbidity and mortality (Krause-Heuer et al., 2009). Notwithstanding the emergence of new chemotherapeutic drugs and novel therapies for cancer, significant challenges remain for understanding tumourigenesis and tumour cell biology, and in developing new, effective strategies for cancer treatment. (Mozafari et al., 2009). Some of these challenges include;

- Identification of new tumour targets
- Drug potency, due to inadequate concentration at the cell surface.
- Non-selective nature of cytotoxic agents and a low therapeutic index.
- Development of multi-drug resistance (MDR)

The development of therapeutic monoclonal antibodies (mAbs) has shown promise in treatment of cancer amongst other indications. At present there are around 30 approved therapeutic mAbs predominantly for treating cancer and diseases associated with inflammation (Walsh, 2010). Anti-cancer antibodies are designed to target tumour cell surface antigens, with subsequent eliciting of an immune response on tumour binding. Most commonly this is termed antibody dependent cellular cytotoxicity (ADCC), and activated natural killer cells are recruited to attack the tumour. Currently approved anti-cancer therapeutic mAbs targeting the tumour cell surface are specific for antigens including EGFR, HER2/*neu*, CD20, CD52 and CD33. Other targets which are receiving much attention are the mucins, principally MUC1 (there is no existing approved antibody for MUC1), IGFR and CEA and cancer stem cell antigen CD44. The future development of monoclonal antibody (and other) cancer therapies will be contingent on the identification, development and validation of new tumour targets. However identification of new tumour biomarkers that reliably and accurately diagnose early stage cancer has not been met with great success. As an example, a group of researchers, as part of the Early Detection Research Network (EDRN), tested a group of recently discovered putative biomarkers for ovarian cancer, however none were superior to CA-125, which has been used extensively for 30 years. Notwithstanding the significant challenges in the discovery and development of clinically useful biomarkers, proteomics will be central for the discovery of new, novel biomarkers for early detection and diagnosis; some of these biomarkers may be suitable for development as novel drug targets (Pastwal et al., 2007).

Recently the International Cancer Genomes Consortium (ICGC) was formed, with a charter to co-ordinate and integrate large-scale cancer genome sequencing projects, focusing on 50 different types of cancers (The International Cancer Genome Consortium, 2010). The expanded studies will consist of investigating around 25,000 specific cancers (biopsy material from individuals). The primary objectives of the consortium were made public in April 2008 and were released in April 2008 (http://www.icgc.org/files/ICGC_April_29_2008. pdf). However these studies will be at the genomic, epigenomic and transcriptomic levels. At the proteomic level, the Human Proteome Organization (HUPO, http://www.hupo.org/) instigated the Human Proteome Project (HPP), a co-ordinated global initiative to map the protein-based molecular architecture of the human body. This initiative, will aid in the discovery and cataloging of new tumour associated antigens and potential targets.

## 2.3 Evolution of methods of analysis using proteomics

The term proteomics, encompassing the analysis of and tools used to examine proteins expressed by a genome was coined only about 15 years ago (Wilkins et al., 1996). The progression and development of proteomics since this time however, has naturally afforded a refinement in techniques and methods for simultaneously detecting, identifying and quantifying proteins in biological samples. Fundamental to any identification or further characterization is the ability to first separate proteins from complex samples. This is particularly important for samples such as blood where it is estimated that the dynamic range of proteins is greater than 10exp9. With such a vast dynamic range, just 22 proteins account for 99% of protein content in blood (reviewed in Simpson et al., 2008). With respect to the search for protein biomarkers, the situation is also complicated by the notion that most biomarkers will be low abundance proteins. As protein function and levels of abundance are often altered in disease states, identification of such changes by comparison of healthy and disease samples will allow a greater understanding of the disease, provide new therapeutic targets, as well as identify markers of disease status. Establishment of what could be defined as the "healthy phenotype", will depend on detailed characterization of the proteomes of healthy and diseased states of cells/tissue. One school of thought suggests that by creating complex "proteomic fingerprints" of healthy and diseased states (and transitions thereof), one may recognize perturbations from the healthy state phenotype before manifestation of the disease state. Therapeutic intervention during the transition-to-disease state may instigate a reversion to the healthy phenotype. Thus a systems biology approach to studying the proteomes of cells in normal and diseased states, and also the network of protein-protein interactions, should enhance the opportunity for attaining this goal.

Proteomics and the tools to identify and quantify proteins have evolved substantially since its conception. The availability of genomic information, particularly the complete human genome sequence has in many ways pushed the bottleneck from the genomic to proteomic arena. Developments include tools for protein separation, protein identification, quantification and automated processes. The following sections provide a summary of some of the major approaches to protein separation, identification and quantification using both gel and gel-free proteomic methods. Separation of proteins can be based on one or more physical or biochemical parameter including size, pI, sub-cellular location, or other depletion / enrichment strategies, with separation involving two or more 'orthogonal' approaches providing greater separation than a single dimension alone.

### 2.3.1 Two-dimensional gel electrophoresis

One of the earliest approaches to protein separation was based on the use of two-dimensional gel electrophoresis (2DGE), in which proteins are first separated by charge in a pH gradient (isoelectric focusing), followed by separation based on size in SDS-PAGE gels (O'Farrell, 1975). The gel is subsequently stained to visualize proteins, with each protein spot-volume representative of abundance of the protein(s) within it. This approach, particularly when combined with other upstream fractionation steps (e.g. sub-cellular) can provide separation and resolution of a large number of proteins (Cordwell et al., 2000) and has been used in various projects aiming to identify differentially expressed proteins – potentially biomarkers, by comparing control vs. diseased samples via gel-analysis software and statistical tests. Using 2DGE, identification of candidate biomarkers has been achieved for a range of diseases including atherothrombotic ischemic stroke (Brea et al., 2008), pancreatic cancer (Park et al., 2011) and breast cancer (Lee, et al., 2011) with many studies identifying new potential markers.

### 2.3.2 Differential In-Gel Electrophoresis (DIGE)

Improvements to 2DGE include the addition of small fluorescent tags (CyDyes) on protein samples prior to separation, thus allowing multiple samples to be combined in the same physical gel (Tonge et al., 2001). This approach, known as Differential In-Gel Electrophoresis or DIGE, circumvents some of the problems of quantifying proteins across different samples in different gels as the distinct fluorescent tags on different samples allow the researcher to detect proteins from control vs. disease samples simultaneously. The inclusion of a pooled internal standard representing a mix of all samples can help circumvent some of the technical difficulties (e.g. gel warping and spot matching) that arise from single-sample gels. An advantage of DIGE is that due to the sensitive fluorescent nature of the CyDye labels low amounts of sample (as little as 10 µg) are required for analysis. However, a caveat is that although protein separation and quantification can be achieved, the identity of the proteins still remains unknown unless the protein spots are excised and further analysed, which can be difficult due to the low amounts of protein used in the analysis. Limitations also exist in the dynamic range of protein detection, estimated at 10exp4 (reviewed in Rabilloud, 2002) Typically a separate, unlabelled gel with larger amounts of protein loaded are required to be run and subsequently cross matched with the original DIGE experiment, as described by Matigian et al., 2010. Despite these difficulties, 2DGE has a proven track record in the separation and identification of proteins, with numerous differentially expressed proteins and potential biomarkers uncovered. 2DGE has been applied to a wide range of sample types including tissues (e.g. breast, skin, brain) and fluids such as cerebrospinal fluid, serum, urine, and tears targeting diseases such as various cancers, Alzheimers disease and dementia, cardiovascular diseases, infections such as HIV, conjunctivitis and toxoplasmosis. Overall, 2DGE with or without fluorescent labeling of samples does provide good separation of proteins, but is time consuming and a laborious process both in the gel procedures, and analysis of 2D protein spot profiles.

## 2.4 Mass spectrometry-based approaches

Mass spectrometry (MS) has formed the basis for standard protein identification for many years, typically in a 'bottom-up' approach (in which proteins are digested, usually with trypsin) and the resulting peptides analyzed to determine protein identity, but also in some cases by 'top-down' approaches where intact proteins form the basis of analysis. MS-based approaches hold some advantages over traditional 2DGE methods in that samples can potentially be analyzed and identified simultaneously through methods such as two dimensional LC-MS/MS using a combination of strong cation exchange followed by reverse phase separation of peptides. In comparison to gel-based approaches, MS analyses appear to be more effective at identifying low abundance proteins, as well as those with extreme physical properties such as molecular weights (low or high) or pI values, which are often difficult to resolve on gels. MS-based analyses also offer better prospects for automation of separation, analysis and identification of proteins.

With the ability to rapidly identify large numbers of proteins via MS, the emphasis has since shifted to also quantifying those proteins detected by MS. Broadly speaking, the approaches to quantifying proteins via mass spectrometry can be based on labeled or label-free methods. Each approach has its own advantages and disadvantages. Label-based methods include Isotope-Coded Affinity Tag or ICAT, and Isotope Tag for Relative and Absolute Quantitation or ITRAQ, in which multiple samples can be labeled, mixed and then analyzed simultaneously via MS to avoid technical issues relating to reproducibility that may be encountered with label-free approaches.

## 2.4.1 ITRAQ, ICAT and SILAC

ICAT and ITRAQ differ in their labeling chemistries and site of attachment. For ICAT, cysteine (cys) residues are targeted and selected for via avidin affinity chromatography. The enrichment of only those cys-containing peptides provides one avenue to quantify samples without the complexity of analyzing all peptides in a sample. However, ICAT becomes problematic for analysis of proteins which lack any cys residues. Furthermore, as reviewed in Patton et al., 2002, approximately 70% of proteins contain four or less cys residues thus limiting the usefulness of this approach.

ITRAQ utilizes lysine residues for labeling (Ross et al., 2004), thus avoiding the problem of limited cys residues encountered with ICAT. ITRAQ is a multiplexed approach, where tags are based on isobaric reagents. This means that up to eight different samples can be labeled with unique tags. The physical properties of the tags differ only in the isotopes used in their synthesis, meaning during LC separations, and in MS mode are identical. Only upon fragmentation in MS/MS mode are the isobaric tags distinguishable. The result is that proteins can be identified via MS/MS and due to unique reporter ions from the ITRAQ tag, the protein can also be quantified. The initial ITRAQ labels were designed to label up to four different samples, although tags to label and detect up to eight different samples are now available. Limitations for ITRAQ lie with the difficulty of identify proteins and quantify them when uniquely expressed in only one sample type eg a protein expressed only in the diseased state. Some technical difficulties, in particular with 8plex tags, resulting in a reduction in identification efficiency have also been reported (Thingholm et al., 2010).

Other label-based strategies such as Stable Isotope Labeling with Amino acids in Cell culture (SILAC) also exist. This method is based on labeling proteins in culture with heavy and light forms of amino acids. The approach is a useful way of comparing two samples, but is limited only to cells grown in culture or in some cases, animal models (Zanivan et al., 2012) and would not be applicable to human or clinical studies due to the use of radioactivity.

### 2.4.2 Selected reaction monitoring / multiple reaction monitoring

Selected reaction monitoring (SRM), also known as Multiple Reaction Monitoring (MRM) provides a targeted approach to quantifying proteins in a sample. It advances the 'global' approach to quantification by simply targeting those proteins specifically of interest to the researcher. Typically MS instruments such as triple quadrupoles are used, where a precursor mass representing a peptide (typically tryptic) from the protein of interest is selected, fragmented and specific product ions unique to that peptide, monitored. Generally for each protein of interest a number of precursor ions, and subsequent product ions are monitored to ensure specificity. This information is then used to identify and quantify the proteins present. The main limitation of SRM/MRM analyses is that it is essential to know the proteins of interest beforehand, so that appropriate precursor / product ion can be monitored.

SRM / MRM is a targeted approach, where identify of the protein(s) of interest must be known beforehand. The usefulness of SRM/MRM analyses is thus as a downstream technology after discovery-phase experiments have concluded, and candidate proteins of interest requiring quantification already identified. The advantage of SRM / MRM assays is the ability to simultaneously monitor numerous potential biomarkers in a single analysis and quantify protein levels and is thus currently a popular area of investigation. To help with SRM / MRM analyses, a consortium called SRMAtlas has been established (www.mrmatlas.org) to quantify proteins in complex samples by MS. As well as human entries, mouse and yeast information is also contained, and provides both web-interface and command line tools to search for assays. This readily available information means researchers can potentially circumvent some method development steps as optimal coordinates for SRM / MRM transitions of numerous proteins are available.

### 2.4.3 Stable isotope standards and capture by anti-peptide antibodies

Stable isotope standards and capture by antipeptide antibodies (SISCAPA) is a method which allows the quantification of peptides from complex digests. Originally described by Anderson et al., 2004 the method utilizes stable-isotope-labeled internal standards for comparison with (unlabelled) peptides that are enriched via anti-peptide antibodies, with subsequent quantification performed by electrospray mass spectrometry. The approach offers increased sensitivity over non-enriched methods, particularly when coupled with SRM / MRM assays. SISCAPA also offers potential in the verification of diagnostic protein panels from large samples as well as increased efficiency in assay time for the bind/elute process over conventional reverse phase separations. There are also distinct advantages over traditional techniques such as ELISA in development time for biomarker assays (Whiteaker et al, 2009). The main disadvantage of SISCAPA is the need for preselected targets, as well as the cost in producing the internal peptide standards and generation of the peptide binding antibodies. However, given the sensitivity of the assay (only low fmol – pmol amounts are required), and the fact the antibody itself can be recycled and used again means on-going costs can be reduced. Since the original design of SISCAPA, refinements in the assay have been developed to reduce loss of low abundance peptides, automated processing steps, and improvement in antibody sources i.e. from polyclonal to monoclonal (Anderson et al., 2009; Schoenherr 2009). As this method is only a relatively recent development, no biomarkers have as yet been published as validated with this approach, although proof-of-principle experiments have been performed with established biomarkers such as tropinin I (cTnI) (Kuhn, et al., 2009) and thus SISCAPA remains a promising tool.

### 2.4.4 Alternative strategies

In addition to the above technologies, other strategies have been developed to complement gel and MS approaches to detect biomarkers through improved sample preparation methods. For example, hexapeptide libraries, based on combinatorial peptide libraries offer a way to deplete samples of highly abundant proteins (Guerrier et al., 2008). In this innovative technique, a large collection of specific hexapeptides (hexapeptide library) is attached to beads. The complex protein sample of interest is mixed with the hexapeptide-bead library. The peptide library is of high diversity and so it would be expected that a specific peptide(s) in the library would have affinity for each individual protein in the complex sample. After separation of the beads from the mixture, the adsorbed proteins are eluted from the beads. As each hexapeptide is equally represented within the library, the end result is that the abundant proteins are depleted, while proteins of low abundance are concentrated. This approach is particularly useful for biological fluids (serum, saliva, urine etc) which have particularly large dynamic range of protein abundance. For example, hexapeptide enrichment of urine has uncovered an additional 251 proteins that were not previously known to be present in this fluid (Castagna et al., 2005). Although depletion/enrichment strategies may not, in their own right, uncover biomarkers, their usefulness lies in the ability to mine deeper into the proteome of these highly complex samples so that low abundance proteins can be identified.

Depletion / enrichment strategies can be problematic if the abundant protein is a carrier for low abundance molecules and the use of depletion strategies must be done with caution. For example it has been shown that the depletion of albumin from human plasma can also remove low abundance proteins such as cytokines (Granger et al., 2005). More recent studies (Bellei et al., 2010) have also concluded that removal of high-abundance proteins can result in a loss of non-targeted, less abundant proteins. Obviously unintentional and unknowing loss of low abundance proteins is a cause for concern in the search for biomarkers of disease.

Other approaches for analysis of samples include Surface Enhanced Laser Desorption Ionisation (SELDI) and Matrix Assisted Laser Desorption Ionisation (MALDI), which have both been utilised particularly in examination of body fluids such as serum for biomarker discovery. This is effectively a "protein pattern recognition" approach (reviewed in Zhan & Desiderio, 2010) which compares profiles from control versus disease samples to identify those proteins differentially expressed. This approach has been used in particular for analysis of cancer patients.

### 2.4.5 Post-translational modifications

The majority of the above technologies focus on protein expression and differential expression in control vs. disease states. However, greater emphasis in the future on protein post-translational modifications (PTMs) such as glycosylation and phosphorylation will be needed. Already, perturbations in modifications of proteins by the glycan O-linked B-N-acetylglucosamine (O-GlcNAc) has been implicated in a range of diseases, including Alzheimers and diabetes (reviewed in Dias & Hart 2007). Similarly, differential phosphorylation has been identified in diseased states such as cancer compared to control patients (Semaan et al., 2011).

### 2.4.6 The Human Proteome Project

Fundamental to rational design for disease treatment and prevention is the understanding of genes present, and the expression and function of gene products, including proteins involved in the disease process. The Human Genome Project (HGP) was established to map all genes encoded in the human genome. Surprisingly, the total number of protein coding genes, only approximately 20,300, was substantially lower than expected, with increased complexity presumably due to splice variants, and post-translational modifications. Following on from this ground-breaking work, is the recent establishment of the Human Proteome Project which aims to map the human proteome (Legrain et al., 2011). At present, of the protein-coding genes in humans identified in the human genome, approximately one third have not been detected at the protein level, while for many others, basic information such as abundance, sub-cellular localization, or function are unknown. Mapping of the proteome will be valuable in understanding human biology, and downstream applications in developing diagnostics, prognostics and new therapies to treat diseases. The HPP will have a 'gene-centric' approach to map information about proteins back to gene loci. HPP will aim to address three parts (HUPO Views, 2010):

- Identification and characterization of proteins from every gene.
- Distribution of proteins in all normal tissues and organs.
- Mapping of pathways and protein networks and interactions.

With respect to sample type, bodily fluids relatively easily attainable, such as urine, saliva, tears as well as those requiring more slightly more invasive methods for collection such as serum, plasma and CFS have all been analyzed for a variety of diseases. Fluids as opposed to solid tissues would generally form a better basis for determining personalized signatures and biomarker detection due to their ease of attainment. There has been some question over whether blood is the best choice for searching for biomarkers. The rationale has been that specific proteins are secreted by the body from different organs, and that these can represent a biological "fingerprint" of physiological state (reviewed in Simpson et al., 2008).

Using model systems such as mice, researchers have shown changes in the plasma proteome prior to any clinical evidence of breast cancer (Pitteri et al., 2011). A separate study in humans (Li, 2011) also suggests that it may be possible to observe proteome plasma changes prior to diagnosis. Plasma proteomics has also been used in the search for pre-diagnostic markers in other diseases such as coronary heart disease (Prentice et al., 2010). The ability to identify proteome changes prior to manifestation of disease phenotypically, will potentially improve patient outcome, particularly for those diseases such as cancer where early diagnosis strongly correlates with survival rates.

## 3. Conclusion

The ability to define proteomic "signatures"' for individuals will vastly enhance the ability of the medical community to diagnose and treat diseases, as well as potentially identify disease before symptoms appear. Early treatment will in turn prolong life, as well as potentially address healthcare costs through the application of more refined and defined therapies suitable for individual patients. The heterogeneity of some diseases such as breast cancer, in which specific proteins, e.g. progesterone receptor, estrogene receptor and HER-2 may or may not be expressed, make it difficult to broadly treat patients, as a 'one size-fits all' approach does not always apply, and emphasizes the need for individualized and personalized medicine. By examining the proteome, it is possible to gain a better understanding of the heterogeneity present in an individual and potentially can help determine best choice of therapies, as well as indicate disease status and progression. Given the complication of genetic factors and environmental influences on an individual, personalized medicine strategies will require complementation of proteomic data with other areas and strategies for analysis and compile this information to determine diagnostic approaches and tailor therapeutic strategies for the individual.

As yet, despite the excitement of biomarker discovery, and the vast number of publications claiming detection of biomarkers for a specific disease, the majority of candidate biomarkers are yet to be validated or used in clinical settings. However, once candidate biomarkers are confirmed, the emphasis will be on high-throughput approaches to expand analyses to greater numbers of samples. Clearly the proteomic tools available to detect and characterize samples, particularly in a high-throughput quantitative fashion are now a reality. Thus, personalized medicine is not far off the horizon. We anticipate a new era of therapeutic approaches and more refined medicinal treatments for diseases which will be more targeted and precise, not just for the disease, but for the individual, based on establishment of "proteomic fingerprints". In addition to greater confidence in diagnoses, proteome signatures would allow a more individualized and targeted approach to therapy. Potentially, such signatures may also provide better insight into future recurrences of the disease.

Besides the quest for discovery, research and development of new and unique biomarkers, other facets of biomarker research incorporate aims such as improving reliability, increasing the speed of detection and reducing the amount of sample needed for analysis. However, the search for biomarkers is particularly important for those diseases such as breast cancer, for which there are no current clinical biomarkers available, and for which mortality is tightly related to disease stage in the initial surgery (Bohm et al., 2011). Using proteomics however, a biomarker signature for non-metastatic breast cancer has been uncovered. This study (Bohm et al., 2011) found using serum samples and SELDI-TOF and MALDI-

TOF/TOF analyses, a combination of 14 biomarkers that can identify breast cancer patients from controls, with a specificity of 67%. It is unlikely at this stage that this panel or signature will entirely replace imaging diagnostics, but does have the potential to aid current diagnostic approaches, particularly when cancer survival rates are greatly improved with early detection, while tumors <5 mm are normally not detected.

One of the fundamental problems of assigning 'biomarker' status for a protein found to be differentially expressed in a disease is the overlap of these differentially expressed proteins across different diseases. A number of proteins have been implicated across a number of different diseases, making the notion of a single biomarker to indicate a specific disease more difficult. For example, serum amyloid A has been proposed as a prognostic marker for melanoma (Findeisen et al., 2009), breast cancer (Schaub et al., 2009), atherothrombotic ischemic stroke (Brea et a., 2009). Potentially, for greater confidence in disease diagnosis or prognosis, it may be required that a suite of biomarkers, be needed to provide greater specificity and confidence.

The significance of the future development of personalized medicine is far reaching, and will allow/facilitate the following:

- Predicting a patient's response to drugs.
- Development of customized' prescriptions.
- Minimizing, or in some cases eliminating adverse events.
- Improving rational drug development.
- Improving drug R&D and the approval of new drugs - better designed clinical trials based on genomic/proteomic information.
- Screening and monitoring certain diseases e.g. advanced diagnosis before disease symptoms.
- Reducing the overall cost of healthcare.

If the concept of routine personalized medicine is to become a reality in the future, the development of new proteomic techniques and methodologies will be vital, and will build on current methodologies now available.

## 4. References

Altman, R. (2007). PharmGKB: a logical home for knowledge relating genotype to drug response phenotype. *Nature Genetics*, Vol.39, pp. 426, ISSN 1061-4036

Anderson, N.L.; Anderson, N.G.; Haines, L.R.; Hardie, D.B.; Olafson, R.W.; & Pearson T.W. (2004). Mass Spectrometric Quantitation of Peptides and Proteins Using Stable Isotope Standards and Capture by Anti-peptide Antibodies (SISCAPA) *Journal of Proteome Research*, Vol.3, pp. 235-244, ISSN 1535-3893

Anderson, N.L.; Jackson, A.; Smith, D.; Hardie, D.; Borchers, C. & Pearson, T.W. (2009) SISCAPA Peptide Enrichment on Magnetic Beads Using an In-line Bead Trap Device. *Molecular & Cellular Proteomics*, Vol.8, pp.995-1005, ISSN 1535-9476

Banerji, U. (2009). Heat shock protein 90 as a drug target: some like it hot. *Clinical Cancer Research*, Vol.15, pp. 9–14, ISSN 1078-0432

Butler, G.S. & Overall, C.M. (2009). Proteomic Identification of Multitasking Proteins in Unexpected Locations Complicates Drug Targeting *Nature Reviews Drug Discovery*, Vol.8, pp. 935-948, ISSN 1474-1776

Bellei, E.; Bergamini, S.; Monari, E.;Fantoni, L.I.; Cuoghi, A.; Ozben, T.; & Tomasi, A. (2010). High-abundance proteins depletion from serum proteomic analysis: concomitant removal of non-targeted proteins. *Amino Acids* Vol.40, pp.145-156 ISSN 0939-4451

Bohm, D.; Keller, K.; Wehrwein, N.; Lebrecht, A.; Schmidt, M.; Kolbl, H. & Grus, F.-H. (2011). Serum proteome profiling of primary breast cancer indicates a specific biomarker profile. *Oncology Reports*, Vol.26, pp. 1051-1056, ISSN 1021-335X

Brea, D.; Sobrino, T.; Blanco, M.; Fraga, M.; Agulla, J.; Rodriguez-Yanez, M.; Rodriguez-Gonzalez, R.; Perez de la Ossa, N.; Leira, R.; Forteza, J.; Davalos, A. & Castillo, J. (2009) Usefulness of haptoglobin and serum amyloid A proteins as biomarkers for atherothrombotic ischemic stroke diagnosis confirmation. *Atherosclerosis*, Vol.205, pp. 561-567, ISSN 0021-9150

Buchen, L. (2011). Missing the Mark: Why Is It So Hard to Find a Test to Predict Cancer *Nature*, Vol.471, pp. 428-432, ISSN 0028-0836

Butler, G.S. & Overall, C.M. (2009). Proteomic Identification of Multitasking Proteins in Unexpected Locations Complicates Drug Targeting *Nature Reviews Drug Discovery*, Vol.8, pp. 935-948, ISSN 1474-1776

Castagna, A.; Cecconi, D.; Sennels, L.; Rappsilber, J.; Guerrier, L.; Fortis, F.; Boschetti, E.; Lomas, L.; & Righetti, P.G. (2005). Exploring the Hidden Human Urinary Proteome via Ligand Library Beads. *Journal of Proteome Research*, Vol.4, pp. 1917-1930, ISSN 1535-3893

Chakravati, A. (2001). To a Future of Genetic Medicine (2001) *Nature*, Vol.409, pp. 822-823, ISSN 0028-0836

Cordwell, S.J.; Nouwens, A.S.; Verrills, N.M.; Basseal, D.J.; & Walsh, B.J (2000). Sub-proteomics based upon protein cellular location and relative solubilities in conjuction with composite two-dimensional electrophoresis gels. *Electrophoresis*, Vol. 21, pp. 1094-1103, ISSN 0173-0835

Dias, W.B. & Hart, G.W. (2007) O-GlcNAc modification in diabetes and Alzheimer's disease. *Molecular Biosystems*, Vol.3, pp. 766-772, ISSN 1742-206X

Filiou, M.; Turck, C. & Martins-de-Souza, M. (2011). Quantitative proteomics for investigating psychiatric disorders. *Proteomics Clinical Applications*, Vol.5, pp. 38-49, ISSN 1862-8354

Granger, J.; Siddiqui, J.; Copeland, S.; & Remick, D. (2005). Albumin depletion of human plasma also removes low abundance proteins including the cytokines. *Proteomics* Vol.5, pp. 4713-4718, ISSN 1615-9853

Guerrier, L.; Righetti, P.G.; & Boschetti, E. (2008). Reduction of Dynamic Protein Concentration Range of Biological Extracts for the Discovery of Low-Abundance Proteins by Means of Hexapeptide Ligand Library. *Nature Protocols*, Vol.3, pp.883-890, ISSN 1750-2799

HUPO Views (2010). A Gene Centric Human Proteome Project. *Molecular and Cellular Proteomics*, Vol.9, pp. 427-429, ISSN 1535-9476

Jain, R. (2004). Role of Pharmacogenomics in the Development of Personalized Medicine *Pharmacogenomics*, Vol.5, pp. 331-336, ISSN 1462-2416

Krause-Heuer, A.M.; Grant, M.P.; Orkey, N. & Aldrich-Wright JR. (2008). Drug Delivery Devices and Targeting Agents for Platinum(ii) Anticancer Complexes. *Australian Journal of Chemistry,* Vol.61, pp. 675-81, ISSN 0004-9425

Krejsa, C.; Rogge, M. & Sadee, W. (2006). Protein therapeutics: new applications for pharmacogenetics. *Nature Reviews Drug Discovery,* Vol.5, pp. 507-521, ISSN 1474-1776

Kuhn, E.; Addona, T.; Keshishian, H.; Burgess, M.; Mani, D.R.; Lee, R.T.; Sabatine, M.S.; Gerszten, R.E.; Carr, S.A. (2009). Developing Multiplexed Assays for Troponin I and Interleukin-33 in Plasma by Peptide Immunoaffinity Enrichment and Targeted Mass Spectrometry. *Clinical Chemistry,* Vol.55, pp. 1108-1117, ISSN 0009-9147

Lee, S.; Terry, D.; Hurst, D.R.; Welch, D.R.; & Sang, Q-X.A. (2011). Protein signatures in human MDA-MB-231 breast cancer cells indicating a more invasive phenotype following knockdown of human endometase / matrilysin-2 by siRNA. *Journal of Cancer,* Vol.2, pp. 165-176, ISSN 1837-9664

Legrain, P.; Aebersold, R.; Archakov, A.; Bairoch A.; Bala, K.; Beretta, L.; Bergeron, J.; Borchers, C.H.; Corthals, G.L.; Costello, C.E.; Deutsch, E.W.; Domon, B.; Hancock, W.; He, F.; Hochstrasser, D.; Marko-Varga, G.; Salekdeh, G.H.; Sechi, S.; Snyder, M.; Srivastava, S.; Uhlén, M.; Wu, C.H.; Yamamoto, T.; Paik, Y.K. & Omenn, G.S. (2011) The human proteome project: current state and future direction. *Molecular and Cellular Proteomics,* Vol.10, M111.009993 ISSN 1535-9476

Li, C.I (2011). Discovery and Validation of Breast Cancer Early Detection Biomarkers in Preclinical Samples. *Hormones and Cancer,* Vol.2, pp. 125-131, ISSN 1868-8497

Marko-Varga, G.; Ogiwara, A.; Nishimura, T.; Kawamura, T.; Fujii, K.; Kawakami, K.; Kyono, Y.; Tu, H.; Anyoji, H.; Kanazawa, M.; Akimoto, S.; Hirano, T.; Tsuboi, M.; Nishio, K.; Hada, S.; Jiang, H.; Fukuoka, M.; Nakata, K.; Nishiwaki, Y.;Kunito, H.; Peers, I.; Harbron, C.; South, M.; Tim Higenbottam, T.; Nyberg, F.; Kudoh, S and I and Kato, H. (2007). Personalized Medicine and Proteomics: Lessons from Non-Small Cell Lung Cancer. *Journal of Proteome Research,* Vol.6, pp. 2925-2935, ISSN 1535-3893

Matigian, N.; Abrahamsen, G.; Sutharsan, R.; Cook, A.; Vitale, A.; Nouwens, A.; Bellette, B.; An, J.; Anderson, M.; Beckhouse, A.; Bennebroek, M.; Cecil, R.; Chalk, A.; Cochrane, J.; Fan, Y.; Féron, F.; McCurdy, R.; McGrath, J.; Murrell, M.; Perry, C.; Raju, J.; Ravishankar, S.; Silburn, P.; Sutherland, G.; Mahler, S.; Mellick G.; Wood, S.; Sue, C.; Wells, C. & Mackay-Sim, A. (2010). Disease-specific, neurosphere-derived cells as models for brain disorders. *Disease Models and Mechanisms,* Vol.3, pp. 785-798, ISSN 1754-8403

Mattick, J.S. (2009). Deconstructing the dogma: a new view of the evolution and genetic programming of complex organisms. *Annals of the New York Academy of Sciences,* Vol.1178, pp. 29-46, ISSN 0077-8923

Mercer, T.R.; Dinger, M.E.; Sunkin, S.M.; Mehler, M.F. & Mattick, J.S. (2009). Long non-coding RNAs: insights into functions. *Nature Reviews Genetics,* Vol.10, pp. 155-159, ISSN 1471-0056

Miki, Y.; Swensen, J.; Shattuck-Eidens, D.; Futreal, P.A.; Harshman, K,; Tavtigian S.; Liu, Q.; Cochran, C.; Bennett, L.M.; Ding, W.; et al., (1994). A strong candidate for the breast and ovarian cancer susceptibility gene BRCA1. *Science*, Vol.266, pp. 66-71, ISSN 0036-8075

Mozafari, M.R.; Pardakhty, A.; Azarmi, S.; Jazayeri, J.A.; Nokhodchi, A. & Omri, A. (2009). Role of nanocarrier systems in cancer nanotherapy. *Journal of Liposome Research,* Vol.19, pp. 310-21, ISSN 0898-2104

O'Farrell, P.H. (1975). High resolution two-dimensional electrphoresis of proteins. *Journal of Biological Chemistry*, Vol.250, pp. 4007-4021

Park, J.Y.; Kim, S.-A.; Chung, J.W.; Bang, S.; Park, S.W.; Paik, Y.K. & Song, S.Y. (2011). Proteomic analysis of pancreatic juice for the identification of biomarkers of pancreatic cancer. *Journal Cancer Research and Clinical Oncology*, Vol.137, pp. 1229-1238, ISSN 0171-5216

Pastwal, E.; Somiari, S.; Czyz, M. & Somiari, R. (2007). Proteomics in human cancer research. *Proteomics Clinical Applications*, Vol.1, pp. 4-17, ISSN 1862-8354

Patterson, S.; Van Eyk, J. & Banks, R. (2010). Report from the Wellcome Trust/EBI "Perspectives in Clinical Proteomics" retreat – A Strategy to Implement Next-Generation Proteomic Analyses to the Clinic for Patient Benefit: Pathway to Translation. *Proteomics Clinical Applications,* Vol.4, pp. 883-887, ISSN 1862-8354

Patton, W.F.; Schulenberg, B. & Steinberg, T.H. (2002) Two-dimensional gel electrophoresis: better than a poke in the ICAT? *Current Opinion in Biotechnology*, Vol.13, pp.321-328, ISSN 0958-1669

Pitteri, S.J,; Kelly-Spratt, K.S.; Gurley, K.E,; Kennedy, J.; Busald Buson, T.; Chin, A.; Wang, H.; Zhang, Q.; Wong, C.-H.; Chodosh, L.A ; Nelson, P.S.; Hanash, S.M.; & Kemp, C.J. (2011). Tumor Microenvironment-Derived Proteins Dominate the Plasma Proteome Response during Breast Cancer Induction and Progression. *Cancer Research,* Vol.71, pp. 5090-5100, ISSN 0008-5472

Prentice, R.L.; Paczesny, S.J.; Aragaki, A.; Amon, L.M.; Chen, L.; Pitteri, S.J.; McIntosh, M.; Wang, P. ; Buson Busald, T,; Hsia, J. ; Jackson, R.D. ; Rossouw, J.E. ; Manson, J.E. ; Johnson, K.; Eaton, C. & Hanash, S.M. (2010). Novel proteins associated with risk for coronary heart disease or stroke among postmenopausal women identified by in-depth plasma proteome profiling. *Genome Medicine*, Vol.2, pp. 48-60, ISSN 1756-994X

Rabilloud, T. (2002). Two-dimensional gel electrophoresis in proteomics : Old, old fashioned, but it still climbs up the mountains. *Proteomics* Vol.2 pp. 3-10, ISSN 1615-9853

Remily-Wood, E.; Liu, R.; Xiang, Y.; Chen, Y.; Thomas, C.; Rajyaguru1, N.; Kaufman, L.; Ochoa, J.; Hazlehurst, L.; Pinilla-Ibarz, J.; Lancet, J.; Zhang, G.; Haura, E.; Shibata, D.; Yeatman, T.; Smalley, K.; Dalton, W.; Huang, E.; Scott, E.; Bloom, G.; Eschrich, S. & Koomen, J. (2011). A database of reaction monitoring mass spectrometry assays for elucidating therapeutic response in cancer. *Proteomics Clinical Applications*, Vol.5, pp. 383-396, ISSN 1862-8354

Ross. P.L.; Huang, Y.N.; Marchese, J.N.; Williamson, B.; Parker, K.; Hattan, S.; Khainovski, N.; Pillai, S.; Dey, S.; Daniels, S.; Purkayastha, S.; Juhasz, P.; Martin, S.; Bartlet-Jones, M.; He, F.; Jacobson, A. & Pappin, D.J. (2004). Multiplexed Protein Quantitation in Saccharomyces cerevisiae using amine-reactive isobaric tagging reagents. *Molecular and Cellular Proteomics,* Vol.3, pp. 1154-1169, ISSN 1535-9476

Schoenherr, R.M.; Zhao, L.; Whiteaker, J.R.; Feng, L.-C.; Li, L.; Liu, L.; Liu, X. & Paulovich, A.G. (2010) Automated Screening of Monoclonal Antibodies for SISCAPA Assays using a Magnetic Bead Processor and Liquid Chromatography-selected Reaction Monitoring-mass Spectrometry. *Journal of Immunological Methods,* Vol.353, pp. 40-61, ISSN 0022-1759

Simpson, R.J.; Berhhard, O.K.; Greening, D.W. & Moritz, R.L. (2008). Proteomics-driven cancer biomarker discovery: looking to the future. *Current Opinion in Chemical Biology,* Vol.12, pp. 72-77, ISSN 1367-5931

Semaan, S.M. ; Wang, X. ; Stewart, P.A.; Marshall, A.G. & Sang, Q.X.A (2011). Differential phosphopeptide expression in a benign breast tissue, and triple-negative primary and metastatic breast cancer tissues from the same African-American woman by LC-LTQ/FT-ICR mass spectrometry. *Biochemical and Biophysical Research Communications,* Vol.412, pp. 127-131, ISSN 0006-291X

Sherman, J.; McKay, M.; Ashman, K. & Molloy, M.P. (2009). Unique Ion Signature Mass Spectrometry, a Deterministic Method to Assign Peptide Identity. *Molecular and Cellular Proteomics,* Vol.8.9, pp. 2051-2062, ISSN 1535-9476

The International Cancer Genome Consortium (2010). International Network of Cancer genome Projects. *Nature,* Vol.464, pp. 993-998, ISSN 0028-0836

Thingholm, T.E.; Palmisano, G.; Kjeldsen, F. and Larsen, M.R. (2010). Undesirable Charge-Enhancement of Isobaric Tagged Phosphopeptides Leads to Reduced Identification Efficiency. *Journal of Proteome Research,* Vol.9, pp. 4045-4052

Tonge, R.; Shaw, J.; Middleton, B.; Rowlinson, R.; Rayner, S.; Young, J,; Pognan, F.; Hawkins, E.; Currie, I. & Davison, M. (2001). Validation and development of fluorescence two-dimensional differential gel electrophoresis proteomics technology. *Proteomics,* Vol.1, pp. 377-396, ISSN 1615-9853

Walsh G. (2010). Biopharmaceutical Benchmarks. *Nature Biotechnology,* Vol.28, pp. 917-924.

Weinshilbaum, R. & Wang, L. (2004). Pharmacogenomics: Bench to Bedside. *Nature Reviews Drug Discovery,* Vol.5, pp. 38-49, ISSN 1474-1776

Whiteaker, J.R.; Zhao, L.; Anderson, L. & Paulovich, A.G. (2010). An Automated and Multiplexed Method for High Throughput Peptide Immunoaffinity Enrichment and Multiple Reaction Monitoring Mass Spectrometry-based Quantification of Protein Biomarkers. *Molecular and Cellular Proteomics,* Vol.9, pp. 184-196, ISSN 1535-9476

Wilkins, M.R.; Pasquali, C.; Appel, R.D,; Ou, K,; Golaz, O.; Sanchez, J.-C.; Yan, J.X.; Gooley, A.A.; Hughes, G.; Humphery-Smith, I,; Williams, K.L. & Hochstrasser, D.F. (1996). From Proteins to Proteomes: Large Scale Protein Identification by Two-Dimensional Electrophoresis and Amino Acid Analysis. *Biotechnology,* Vol.14, pp. 61-65, ISSN 0733-222X

Wooster, R.; Bignell, G.;, Lancaster J, Swift S, Seal S, Mangion J, Collins N, Gregory S, Gumbs C, Micklem G. et al., (1995). Identification of the breast cancer susceptibility gene BRCA2. *Nature,* Vol.375, pp. 789-792, ISSN 0028-0836

Zanivan, S.; Krueger, M. & Mann, M. (2012). In vivo Quantitative Proteomics: The SILAC mouse. *Methods in Molecular Biology,* Vol.757, pp. 435-450, ISSN 1064-3745

Zhan, X. & Desiderio, D.M. (2010). The use of variations in proteomes to predict, prevent, and personalize treatment for clinically nonfunctional pituitary adenomas. *EPMA Journal*, Vol.1, pp. 439-459, ISSN 1878-5077

# Strategies for Protein Separation

Fernanda Salvato[1], Mayra Costa da Cruz Gallo de Carvalho[2]
and Aline de Lima Leite[3]
[1]*Universidade de São Paulo, Escola Superior de Agricultura Luiz de Queiroz*
[2]*Empresa Brasileira de Pesquisa Agropecuária (EMBRAPA)*
[3]*Universidade de São Paulo, Faculdade de Odontologia de Bauru*
*Brazil*

## 1. Introduction

The proteome of a cell or tissue depends on cellular and environmental conditions, showing a dynamic system subject to large variations. To study these large changes of variability and quantity, proteomics has emerged, providing techniques dedicated to global characterization of all proteins simultaneously. The expectation is that this information will produce new insights into the biological function of proteins in different physiological states of a cell or tissue.

The proteome has a dynamic and complex nature that is the result of many post-translational modifications, molecular interactions, and a variety of proteins arising from alternative mRNA splicing. With this in mind, the number of modified and unmodified proteins found in any biological system is much bigger than the number of genes (Anderson et al., 2004), which is why mRNA expression may not correlate with protein content (Rogers et. al, 2008). In addition, not all proteins are expressed in the same or similar level in the proteome. For example, the enzyme Rubisco comprises 30–50% of leaf proteome (Feller et al., 2008), which is a big issue in the proteomic assessment of low-abundance proteins. In fact, the majority of proteins are in the low-abundance level. To overcome these challenges, the proteome must be fractionated for effective detection and quantification by mass spectrometry (MS). Consequently, the analysis of proteins on the large or small scale is dependent on separation methods.

As the ultimate goal in proteomics is to resolve all individual proteins in the cell, although it is quite difficult to find a separation method that could accommodate the diversity of proteins equally, protein separation methods directly affect the achievement of reliable results. Such methods are based on the physical or chemical properties of different proteins, such as their mass or net charge.

The combination of sequential methods exploiting different properties can provide high-resolution analysis of very complex protein mixtures. Then, current analytical strategies can reach different levels of resolution depending on the platform used. Two-dimensional gel electrophoresis (2DGE) and multidimensional liquid chromatography (MDLC) are the two methods that dominate the separation steps in proteomics. The differences of each strategy are basically related to sensitivity, automation, and high-throughput possibilities. In this chapter, the limitations and principles of these techniques will be discussed.

## 2. 2D-PAGE: Principles, advantages and limitations

The 2D-PAGE (two dimensional polyacrylamide gel electrophoresis) was developed by Patrick H. O'Farrell who successfully combined two known electrophoresis methods, isoeletric focusing (IEF) and sodium dodecyl sulfate electrophoresis (SDS-PAGE) (O'Farrell, 1975) with the objective of resolving more complex proteomes. The author was brilliant in his idea of combining both techniques once now proteins could be separated by two non-related properties given a uniform distribution throughout the gel. Surprisingly, the paper "High Resolution Two-dimensional Electrophoresis of Proteins" was firstly rejected by the JBC (Journal of Biological Chemistry) journal because of its "speculative character", as pointed out in the commemorative issue of the JBC (2006), but the power of 2D gel electrophoresis in resolving proteomes had already spread rapidly. Although the combination made by O'Farrell had immediately caused great impact on proteins separation, its commercial application in proteomics become possible only after a technical modification that made the 2D gel electrophoresis reproducible. In the mid1980s, was introduced to the 2D-PAGE system, commercial strips with immobilized pH gradients (IPG strip) and instruments for IEF (isoelectric focusing) (Bjellqvist et al., 1982) and, since then, the 2D-PAGE assume a central role in proteomics. Together, the 2D-PAGE and mass spectrometric techniques provided the characterization of thousands of proteins in single gels.

### 2.1 Principles of 2D-PAGE

To perform proteins separation, the two dimensional electrophoresis uses sequentially two non-related physical proprieties. In a first dimension, proteins are separated owing their migration in an immobilized pH gradient. Then, in a second dimension, proteins that occasionally took the same migration point after the first separation could now be separate in the polyacrylamide gel, according to their molecular weight, what guarantees to this technique a greater resolution power than achieved in one dimensional electrophoresis. Protein separation can be achieved as low as 0.1 isoelectric point (pI) unit and 1 kDa in molecular weight (MW) (Figeys, 2005). The spots visualized in a second dimension gel are unique proteins or simple mixture of proteins depending on certain factors that can influence technique resolution. To improve resolution, proteins should be completely denatured, reduced, disaggregated from protein complex and solubilized to disrupt macromolecular interactions (Chevalier, 2010).

In the 2D-PAGE protocol, preparation of protein samples is a fundamental and determining stage in electrophoresis efficiency. Usually, to solubilize samples, buffers containing chaotropic agents (urea and/or thiourea), nonionic or zwitterionic (CHAPS or Triton X-100) detergents, reducing agent (DTT) and proteases and phosphatase inhibitors are used. The chaotropic agents will act in the non-covalent macromolecular interactions, interfering in hydrophobic interactions; surfactants (CHAPS and Triton X-100) will act synergistically with chaotropics preventing the adsorption or aggregation of hydrophobic proteins which after the action of thiourea will have their hydrophobic domains exposed; the reducing agent will reduce protein disulfides breaking up intra and inter molecular interactions and proteases and phosphatase inhibitors will avoid modifications in the proteome. Also to optimize proteases and phosphatase inhibition, diluted TCA or TCA-acetone can be used in the solubilization process. One important aspect in solubilization process is to avoid salts accumulation through dialysis or precipitation. Salts can migrate through the pH gradient in the IPG strip and accumulate in the ends, inducing water accumulation and electric current reduction, what interferes in the focalization process.

The first dimension in 2D-PAGE, also called isoelectric focusing (IEF), is performed in acrylamide gel strips with immobilized pH gradient (IPG strips). The gel in the strip is formed through acrylamide polymerization with amphoteric acrylamide monomers named immobilins. Immobilins with different pKa are added to the acrylamide mixture and after gel polymerization; immobilins are immobilized in the strip generating the pH gradient, that's why strips used in IEF are called immobilized pH gradient or IPG strips. The IPG strips are commercially available in many pH ranges such as 6–9, 6–11 or 7–10. They are sold dried and should be rehydrated to be used. In this process, the rehydration solution must be composed by a commercial mixture of carrier ampholytes containing molecules corresponding to all pIs (isoelectric points) in the strip pH range and by the solubilized protein sample to be separated. Ampholytes act as good buffering agents next to their pIs, assisting proteins in the mixture to migrate in the gel.

Isoelectric focusing like the whole electrophoresis process is based on the migration of charged biomolecules under an electric field. The separation of a protein mixture in a pH gradient occurs because proteins are amphoteric molecules and thus can present negative or positive charges in their ionized groups depending on the pH medium. When an electric current is applied, proteins migrate in the gel while the balance between their charges is positive or negative until the difference between charges became equal to zero (isoelectric point – pI), in this point protein migration ceases and protein get focused. Proteins positively charged, i. e., the ones those are in the strip region where pH value is lower than their pI, keep migrating directly to the positive pole until reach their pI. In the other side, proteins negatively charged, i.e., the ones those are in the strip region where pH value is higher than their pI, keep migrating directly to the negative pole until reach their pI. Focusing process can last from 12 to 20 hours.

After IEF ends, the strip containing focused proteins must be equilibrated with the anionic detergent sodium dodecyl sulfate (SDS) solution that denatured these proteins and forms negatively charged protein/SDS complex. The amount of SDS linked into the protein should be directly proportional to its weight, thus proteins that are totally coupled to SDS will migrate in polyacrylamide gel (SDS-PAGE) only due to their weight. Other reagents in the reaction include Tris-HCl buffer, urea, glycerol, DTT, iodoacetamide and bromophenol blue. The second dimension is performed by placing the IPG strip above and in direct contact with the gel in a system composed by two spaced glass. An electric current is applied and proteins migrate from the strip to a second dimension where they are solved due to their molecular weight. In the second dimension, gel can be heterogeneous: with a superior phase or stacking gel with acrylamide 6% and with an inferior phase or resolution gel containing 12 to 15%. In some cases, gel can be homogeneous with acrylamide 13% (Görg et al., 2000). The second dimension can be performed vertically or horizontally, but only the horizontal systems allow multiple runs simultaneously. Gels usually run with 1 or 2 W of current in the first hour, followed by 15 mA/gel overnight with temperature regulation (10°C to 18°C) (Chevalier, 2010).

To visualize the spots in the gel dyes visible to naked eye or fluorescent dyes can be used. In both cases, are necessary to fix the gel after the run, using an acid (phosphoric acid or acetic acid) or an alcoholic (ethanol or methanol) solution depending on the chosen dyeing protocol (Görg et al., 2000). Among the non-fluorescent dyes are Coomassie Brilliant Blue, Colloidal Coomassie Blue and silver nitrate which detect spots respectively with minimal protein of 50,

10 and 0.5 ng (Patton, 2002 & Smejkal, 2004). Usually, fluorescent dyes used are SYPRO dyes, Flamingo and Deep Purple. All these three dyes are sensitive enough to detect spots with up to 1 ng protein (Patton, 2002) however, because of their high costs, they are less used.

Once stained, gels are scanned and gel image can be analyzed using specific software available. It's always recommended to reproduce the proteome of the sample in at least three gels, representing identical technical repetitions. Software will search for a representative spots profile among all repetitions and, if desirable, compare this generated profile with others previously obtained. To perform the comparison, normally markers spots are designated in the gel and the position of all others spots is determinate using these spots as reference. It's also possible to estimate the volume of interested spots assigning them relative quantification values when the objective is to compare proteins differentially expressed. Through the use of these tools a proteomic map for a determined sample can be assembled and yet information on protein differential expression can be obtained. Among the software available are: Image Master, Progenesis, PDQuest, Samespots, the Melanie package from the Swiss Institute of Bioinformatics, the Phoretix 2D software from Phoretix and Gellab II from Scanalytics.

## 2.2 Why to use 2D-PAGE

The 2D-PAGE cannot be used alone to directly identity proteins through the visualization of resulting spots in the gel, even when proteomic maps and sequence information are available to the tested sample. That's because there's a great variation in the proteome of two identical samples, which beings in the protein extraction technique and solubilization and ends in the electrophoresis acrylamide gel. Thus, the identification of proteins in the spots depends on a sequencing stage performed through mass spectrometry (MS or MS/MS). This workflow is usually assumed in proteomics laboratories once can be easily conducted, is applied in many laboratories despite the structure and offers a resolution power enough to detect hundreds of proteins in one gel. Besides that, the 2D-PAGE system is unique about the possibility to visualize the protein profile of a studied sample, allowing immediate comparison with distinct profiles, interesting spots isolation to further studies or yet the enrichment of labeled proteins or specially stained. All these characteristics guarantee its massive application in proteomics characterization. Many other high throughput "gel free" strategies to perform protein separation are available today, but the 2D-PAGE system keep being an important toll in different workflows proposed to protein studies.

The proteomic map assembly is, until the present moment, realized merely by two-dimensional electrophoresis. Gel images are digitalized and made available in data banks what enables the *in silico* comparison between different profiles and the selection of interesting spots. The Japanese Bank containing rice proteomic maps shows for example, more than 13000 characterized spots to different tissues and development phases (http://gene64.dna.affrc.go.jp/RPD/). In humans, there are a great number of studies that report the generation of proteomic maps directed to protein identification that works as biomarkers to reproductive dysfunctions and tumor development (Guo et al., 2010 & Klein-Scory et al., 2010).

Another important contribution of 2D-PAGE system to proteomics is found in the identification and relative quantification of differentially expressed proteins between samples, i.e., differential-display proteomics. Until 1997, this assignment was not easy due to proteome variation in identical samples and gel-to-gel variation frequently observed in the repetition of runs from the same sample. In that time, it was necessary to obtain a great number of gels to reach the required reproducibility in the "average gel" and then perform the comparison between samples. Since DIGE technique or difference gel electrophoresis (Unlu et al., 1997) was developed, the reproducibility problem of 2D-PAGE gel was bypassed. The DIGE system consists in a modification in the conventional protocol of 2D-PAGE that make possible to analyze in a unique gel three different samples giving the electrophoresis system a "multiplex" character. Samples are pre-labeled with fluorescent markers such as Cy2, Cy3 and Cy5, pooled and separate in a single run. Therefore, in addition to solve the 2D-PAGE reproducibility problem, DIGE system allow the direct quantification of spots from different samples resolved in the same gel and is much more sensitive due to fluorescent dye labeling raising the gel resolution dynamic range up to 1,000 times (Chevalier et al., 2010).

The third sample used in the DIGE system is an internal running control composed by identical aliquots from each experimental sample. The mixture: internal standard and sample 01 and 02 are labeled, pooled and resolved in the same gel what avoid diversion on sample preparation. The internal standard control is normally labeled with Cy2 dye and the other samples with Cy3 and Cy5 dyes. The quantification of each protein is obtained from the signal Cy3:Cy2 and Cy5:Cy2 ratio. The Cy3:Cy2 and Cy5:Cy2 ratios for each protein are then normalized across all the gels in a large experiment, using the Cy2 signals for separate normalization of each protein under survey (Lilley & Friedman, 2004).

A problem in DIGE lies in the hydrophobicity of the cyanine dyes, which label the protein by reacting to a large extent, with surface-exposed lysines in the protein, and lead to removal of multiple charges from the protein. Consequently, this decreases the solubility of the labeled protein, and in some cases may lead to protein precipitation prior to gel electrophoresis. To address this problem, minimal labeling is generally employed in DIGE. In this reaction only 1-5% of total lysines in a given protein are labeled avoiding protein precipitation. Alternatively to minimal labeling the saturation labeling method can be done for Cy3 and Cy5 dyes by reacting to free cysteines in a protein (Shaw et al., 2003). This strategy circumvents the sensitivity problem of minimal labeling but limits the proteome analysis to proteins that show free-cystein residues (Chevalier, 2010).

The 2D-PAGE can also be very useful to identify post-translation modifications (PTMs). The affinity chromatography systems are usually used to enrich samples containing a specific PTM, but the 2D-PAGE visual character enable the direct selection on spots differentially expressed to a specific PTM. Proteins resolved in gel can be for example, specifically labeled to detect phosphorylations or glycosylations, and after visually selected; proteins are excised and identified using mass chromatography.

## 2.3 Limitations of 2D-PAGE

Technical characteristics related to gel reproducibility and others that can prevent or influence protein resolution are considered the main limitation of 2D-PAGE technique. However, gel reproducibility in 2D-PAGE method is also strongly influenced by own sample biology, what cannot be considered a limitation from the system *per se*. Some limitations associated to 2D-PAGE are pointed and discussed below.

### 2.3.1 Reproducibility

Protocols available for protein extraction can be applied to various types of biological samples, but the efficiency varies a lot depending on the biological characteristics of the sample. It's much simpler to reproduce the proteome from samples with unique cellular types, like a cell culture for example, than from samples containing many distinct types of cells or cells in different development phases, like for example, from an onion root. The cellular type also offers challenges to protein extraction and solubilization procedures, with a higher reproducibility to animal cell than to plant cells, which are cover by cellular walls and are rich in membranous compartments (plastids). Besides these intrinsic factors associated to sample biology, the proteome dynamism represents an important variation source in experimental repetitions from a same sample, especially when the objective is to perform a protein relative quantification. The proteome can be promptly modify by degradative pathways or by any of the hundreds post-translational modification that exists. Furthermore, small variations due to differences in the genetic backgrounds between sample repetitions can introduce relevant variations in the proteome.

Another factor that can reduce importantly gel reproducibility in 2D-PAGE is gel-to-gel variations which begin in the sample preparation and extend to focusing process and SDS-PAGE. Even in simultaneous runs that preserve exactly the same experimental conditions the gel-to-gel oscillations are present. To minimize or eliminate this effect, two alternatives are available: built an average gel from at least three replicates or use a multiplex run system (DIGE).

It's also important to emphasize that the reproducibility problem of 2D-PAGE system can restrict the applicability of proteomic maps databank that are being generated when no sequencing information to interesting spots are available.

### 2.3.2 Resolution

#### 2.3.2.1 Proteins with high molecular weight

Proteins with molecular weight higher than 250kDa cannot be resolved in polyacrylamide gels. To realize this, the ideal is to use an agarose gel followed by an isoelectric focusing (Yokoyama et al., 2009).

#### 2.3.2.2 Low abundance or rare proteins

Low abundance proteins operate in cellular activities of high interest, participating in signal reception, gene activity regulation and in signal transduction cascades. The detection of these proteins is masked in 2D gels by abundant proteins, which depending on the sample can be present in a magnitude concentration up to 12 times higher. That's the case for example, of albumin protein present in plasma samples. One possible strategy to avoid this problem is the depletion of abundant protein through methods such as affinity chromatography (Greenough et al., 2004). This is normally used to plasma sample, but is not yet possible to other many systems. In plant cells, the abundance of ribulose bisphosphate carboxylase/oxygenase (RuBisCo) enzyme mask low abundant proteins and the current used strategy to understate this effect is to reduce sample complexity through the use of IPG strips with overlapping narrow pH ranges (Görg et al., 2000).

#### 2.3.2.3 Hydrophobic proteins and membrane proteins

In 1998, an important paper was published (Wilkins et al., 1998) in which was demonstrated that hydrophobic proteins were almost absent in 2D gels done using urea as the only chaotropic agent in the protein solubilization solution. This information was very valuable once hydrophobic proteins comprise the proteins present in cellular membranes and represent around 30% of total proteome (Molloy, 2000). After this observation, proteins solubilization began to be realized using a combination of higher concentration of urea and a lower concentration of thiourea, a chaotropic agent much more efficient than urea. The elevated concentration of urea was necessary to solubilize thiourea, which was used in lower concentration, because if in higher concentration, thiourea can interfere in protein focusing process (Molloy, 2000). This modification in the original protocol of protein solubilization resulted in a greater efficiency to solubilize hydrophobic proteins but yet the combination of urea-thiourea cannot keep the proteins in solubilized forms in the aqueous environment necessary to IEF. Other variations in the solubilization protocols combining urea and others nonionic or zwitterionic detergents were suggested, but all resulted in a additional identification of only some membrane proteins spots (reviewed by Rabilloud et al., 2008). It was clear that the solubilization of membrane proteins, mainly those with high hydrophobicity (multiple transmembrane domain), could not be achieved under IEF compatible conditions (reviewed by Tan et al., 2008). The gel systems intent to resolve membrane proteins should use strong detergents for solubilization of this kind of proteins and agents that can add charges to the proteins preventing their aggregation. Such gel-based systems (blue native-PAGE or BN-PAGE, clear-native-PAGE or CN-PAGE, benzyldimethyl-n-hexadecylammonium chloride or BAC, and SDS/SDS or dSDS-PAGE) exclude the IEF resulting in a severely impaired gel resolution (reviewed byTan et al., 2008). The resulting spots are generally composed by a misture of proteins that carry different post-translational modification and/or by complexes of membrane and soluble associated proteins (Rabilloud et al., 2008). Other strategies to detect membrane proteins are available using gel free systems, sample pre-fractionation through subcellular fractionation or affinity purification, and the avidin–biotin technology (Elia, 2008). However, there is still a great necessity of development of protocols that allow the high resolution detection of membrane proteins and the simultaneously detection of membrane and soluble proteins. This is especially important when we are looking for desease responses or physiological phenomena because membrane proteins play key functions in normal development, participating in cellular recognition and signal transduction. Identification of altered membrane proteins could lead to the discovery of novel biomarkers in the disease diagnosis (Adam et al., 2002 & Jang & Hanash, 2003) and targets to therapeutic approaches (Bianco et al., 2006).

### 2.3.2.4 Basic proteins

The basic proteins represent approximately one third to half of total cellular proteome. Among them, are ribosomal proteins and nucleases which exhibit pI superior to 10 and because of this reason are poorly resolved in pH ranges available for alkaline proteins (pH ranges 6–9, 6–11 or 7–10). This 2D-PAGE limitation began to be settle with the commercialization of IPG strips comprising pH ranges of 3-12, 6-12 and 9-12 which are successfully used in the resolution of strongly alkaline protein, with pI superior to 11 (Drews et al., 2004 & Görg et al., 1997).

## 3. Principles of liquid chromatography in proteomics

Chromatographic separation methods have been applied in different laboratories around the world to decipher the many complex problems in industry and science, involving, for example, amino acids and proteins, nucleic acids, carbohydrates, drugs, pesticides, etc. This method separates the components of a mixture by the distribution of these components into two phases, where an immiscible stationary phase remains fixed while the other moves through it. The sample components more strongly connected to the stationary phase move very slowly in mobile phase flow, while those linked more weakly to the stationary phase move more quickly. This process results in differential migration of these components.

The main criteria for classification of chromatographic separation are related to the separation mechanism involved and the different types of stages used. Thus, the physical form of the system classifies the general technique as planar or column chromatography. In the former, the stationary phase is prepared on a flat surface, while in the latter, the stationary phase is arranged in a cylinder. In Gas Chromatography (GC), the mobile phase is an inert gas that does not contribute to  the separation process, whereas in Liquid Chromatography (LC), the mobile phase is a liquid that can interact with the solutes, so their composition is very important in the separation process. Supercritical Fluid Chromatography (SFC) utilizes a substance with temperature and pressure higher than the critical temperature (Tc) and critical pressure (Pc) proper to fluids, with the advantage of having lower viscosity than the liquid while maintaining the properties of interaction with the solutes (Skoog et al., 2006).

The LC techniques may be further divided into classic liquid chromatography (LC) and High-Performance Liquid Chromatography (HPLC). LC utilizes glass columns at atmospheric pressure, and the flow rate is due to gravitational forces. HPLC is the automation of LC under conditions that provide for enhanced separations during a shorter time. It utilizes a metal column and the mobile phase flow rate is due to a high-pressure pump, which increases the efficiency achieved in the separation of compounds, thus making HPLC one of the main techniques used in the separation of proteins and peptides from a wide variety of synthetic or biological sources.

The HPLC equipment comprises a reservoir of mobile phase, which contains the solvents used as the mobile phase to achieve selectivity in HPLC; a pumping system; sample injector; columns; and detectors (Figure 1). The pumping system is required to pump the mobile phase and overcome the pressure exerted by the particles of the column. The major requirements for an efficient pumping system include the ability to generate pressure to

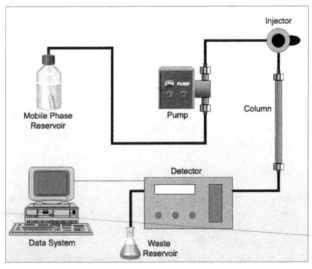

Fig. 1. Scheme of a HPLC system.

6000 psi, with no pulse output; flow rate ranging from 0.1 to 10 mL/min; constant solvent flow (with no variations greater than 0.5%); and corrosion-resistant components. There are two main pumps for HPLC: reciprocal pumps, which are employed in 90% of HPLC systems, consist of a small chamber in which the solvent is pumped by an oscillatory movement of the piston controlled by a motor. Because of this, the flow is not continuous, requiring a shock pulse. Syringe pumps consist of a large chamber equipped with a plunger that is activated by a screw mechanism. The rotation of the screw provides a continuous movement of the mobile phase that is free from pulsations from 0.1 to 5 mL/min. The most common injection system is sampling loops, which allow the introduction of samples up to 7000 psi with excellent precision. These loops can be manual or automated (Oliver, 1991; Meyer, 2010).

In a chromatograph, there are two types of columns: a guard column and separation column. The guard column has a length of 2 to 5 cm and is placed between the injector and separation column, allowing it to retain possible solids that can block the filters of the column and, in some cases, retain materials that can precipitate chemical reactions in the stationary phase. The separation columns (stationary phase) are the heart of a chromatograph, since they are responsible for the separation of the components present in the sample. They consist of a tube of inert material, usually stainless steel, and uniform internal diameter (i.d.), capable of resisting high pressures. They can be classified according to i.d (Saito et al., 2004).

| Column designation | Internal Diameter [mm] |
| --- | --- |
| Conventional HPLC | 3 – 5 |
| Narrow-bore HPLC | 2 |
| Micro LC | 0.5 – 1 |
| Capillary LC | 0.1 – 0.5 |
| Nano LC | 0.01 – 0.1 |
| Open tubular LC | 0.005 – 0.05 |

Table 1. Classification of liquid chromatography according to internal diameter (ID) of columns (Saito et al., 2004)

Silica is the most common stationary phase in HPLC because of advantages such as resistance to high pressures and physicochemical properties. Despite these advantages, silica has two limitations: the first restricts its use in a pH range of 2 to 8 because at pH below 2, bonds of Si-O-Si, which compose the silica and are responsible for maintaining the organic groups immobilized on the silica surface, become more susceptible to hydrolysis. On the other hand, at pH above 8, the hydroxyl groups (OH-) can easily react with the residual silanols, promoting silica dissolution that result in low efficiency and peak enlargement. The second limitation refers to the presence of residual silanol groups that can result on the asymmetry of the peak when basic samples are analyzed (Neue, 1997; Oliver, 1991; Meyer, 2010).

The particle structures are classified as porous, non-porous, and pellicular. The porous particles are most often used for HPLC, since it allows for greater surface area for interactions between the stationary phase and the analyte. Non-porous particles allow faster chromatography without losing efficiency because there is no diffusion of the analyte inside of the particles. However, to keep the sample capacity, it is necessary to use particles with diameters of 1 to 2 μm, as the capacity is 50 times less than that of porous particles. Pellicular particles are constituted of a solid nucleus coated with a thin layer (1–3 μm) of the stationary phase, and they have a good efficiency when analyzing macromolecules due to the fast mass transfer kinetics. The particle shape may be regular (spherical), irregular, or monolithic. The columns packed with spherical particles have a higher resistance to high pressures and good efficiency. The columns packed with irregular particles can have good efficiency when compared to regular particles; however, they have no mechanic stability and can result in higher pressures in the system (Meyer, 2010). Recently, monolithic particles have been introduced in HPLC. They are single pieces of porous silica or a highly intercrossed porous polymer such as polyacrylamide. The skeletons of monolithic particles contain macropores with diameters of approximately 2 μm and mesopores with diameters of approximately 13 nm. Because of those characteristics, they can provide higher flow rates without increasing the pressure, as well as great chemical stability and high permeability (Neue, 1997; Meyer, 2010).

In HPLC, there are different ways to detect the compounds eluting from the column. The ideal detectors are linear, selective and non-destructive and have adequate sensitivity, good stability and reproducibility, and a short response time. However, there are no detectors with all the features mentioned above, so the choice of the detector should be based on objective analysis as well as the type of sample to be analyzed. Liquid chromatographic detectors are basically of two types. Bulk property detectors respond to mobile-phase properties, such as refractive index, dielectric constant, or density. In contrast, solute property detectors respond to properties of solutes, such as UV absorbance, fluorescence, or diffusion current, which are not present in the mobile phase (Skoog et al., 2006; Meyer, 2010). Table 2 shows the major detectors used in HPLC.

| Type of Detector | Limit of Detection | Commercial Available |
|---|---|---|
| Absorbance | 10 pg | Yes |
| Conductivity | 100 pg – 1 ng | Yes |
| Electrochemical | 100 pg | Yes |
| Element Selective | 1 ng | No |
| Fluorescence | 10 fg | Yes |
| FTIR | 1 μg | Yes |
| Light Scattering | 1 μg | Yes |
| Mass Spectrometers | < 1 pg | Yes |
| Optical Activity | 1 ng | No |
| Photoionization | < 1 pg | No |
| Refractive Index | 1 ng | Yes |

Source: Skoog et al., 2006

Table 2. The most common detectors used in HPLC.

High-performance chromatography supplanted gas phase chromatography because it is more versatile; it is not limited to volatile and thermally stable samples, thus allowing a wide choice of mobile and stationary phases. Because the mobile phase carries the solutes through the stationary phase, the correct choice of mobile phase is extremely important in the separation process, as it can completely change the selectivity of separations. The solvents used must be compatible with the stationary phase and detector and the high power of sample solubilization. The elution mode can be isocratic or gradient. In isocratic elution, the separation employs a single solvent or solvent mixture of constant composition, and the mobile phase remains constant with time. Gradient elution, in contrast, utilizes two or more solvent systems that differ significantly in polarity. In this case, when the elution process is begun, the ratio of the solvents varies with time, and separation efficiency is greatly enhanced by gradient elution (Skoog et all., 2006).

The major separation modes that are used to separate most compounds are normal-phase chromatography (NP), reverse-phase chromatography (RP), size-exclusion chromatography (SEC), ion-exchange chromatography (IEX), and affinity chromatography (AC).

In **normal-phase chromatography**, the stationary phase is polar while the mobile phase is non-polar. The retention of analytes occurs by the interaction of the stationary phase's polar functional groups with the polar groups on the particles' surfaces, and they elute from the column by addition of the low polarity compound followed by other compounds of increasing polarity (Figure 2). This method is widely used to separate analytes with low to intermediate polarity (Skoog et al., 2006).

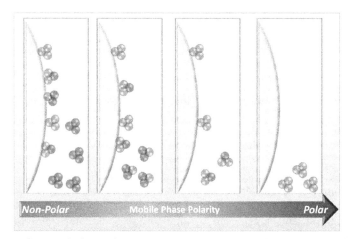

Fig. 2. Diagram of normal-phase chromatography separation. The stationary phase is polar and retains the polar molecule (blue) most strongly. The relatively non-polar molecules (red circles) are quickly eluted by the mobile phase, a non-polar solvent. An increase in mobile phase polarity will move polar molecules through the column.

**Reverse-phase liquid chromatography** has become a powerful tool widely used in the analysis and purification of biomolecules because of the high resolution provided by the technique. It is considered a very versatile technique because it can be used for non-polar, polar, ionizable, and ionic molecules. In RP-HPLC, the separation principle is based on the hydrophobic interaction between the analytes and non-polar groups bound on the stationary phase. Silica is the most common material used for column packing, which consists mainly of silicon dioxide ($SiO_2$) and has octadecyl (hydrocarbons having 18 carbon atoms) and octyl (hydrocarbons having 8 carbon atoms) groups chemically bound to the surface. The mobile phase composition is usually water or a water-miscible organic solvent (methanol, acetonitrile). The analytes adsorbed on the hydrophobic surface remain bound until the higher concentration of the organic solvent promotes the desorption of the molecules from the hydrophobic surface (Figure 3). More hydrophobic analytes are eluted slower than are the hydrophilic analytes (Skoog et al., 2006; GE Healthcare, 2006).

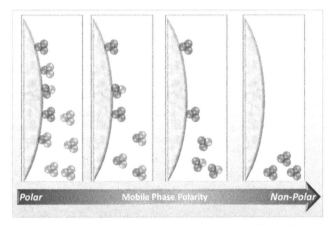

Fig. 3. Diagram of Reverse-phase chromatography separation. The stationary phase is non-polar and retains the non-polar molecule (red) most strongly. The relatively polar molecules (blue circles) are quickly eluted by the mobile phase, a polar solvent. A decrease in mobile phase polarity will move non-polar molecules through the column.

**Affinity chromatography** is the most specific chromatographic method. The separation is based on specific biochemical interactions such as enzyme-inhibitor, antigen-antibody or hormone-carrier. The stationary phase involves an inert matrix coupled with an *affinity ligand* specific for a binding site on the target molecule. The substance to be purified is specifically and reversibly adsorbed to a ligand, immobilized by a covalent bond to a chromatographic matrix. The samples are loaded in an affinity column containing the specific ligand, and the analyte of interest is adsorbed from the sample, while the molecules which have no affinity for the ligand pass through the column (Figure 4). Recovery of molecules of interest can be achieved by changing experimental conditions such as pH values, temperature, or ionic strength or by adding a stronger ligand to the mobile phase. For success in affinity chromatography, some important points have to be considered, such as finding a ligand specific enough and determining the ideal conditions for safe binding between analyte and ligand, as well as the ideal conditions for the retention and elution of the molecules involved (Skoog et al., 2006; GE Healthcare, 2007; Hage, 1999).

**Ion-exchange chromatography (IEC)** is based on the charge properties of the molecules. A stationary phase matrix constituted from a porous and inert material contains charged groups that interact with analyte ions of opposite charge. If these groups are acidic in nature, they interact with positively charged analytes and are called cation exchangers; however, if these groups are basic in nature, they interact with negatively charged molecules and are called anion exchangers. As the matrix material, they can be classified as organic (most common) and inorganic, natural or synthetic. Charged groups binding to the matrix

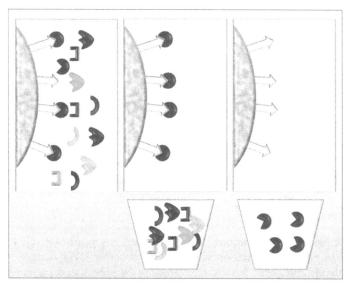

Fig. 4. Affinity chromatography column. The sample is loaded under ideal binding conditions. The target molecules bind specifically to the affinity ligands, while all other sample components, are not adsorbed.

are classified as strong ion exchangers that are completely ionized at a wide range of pH levels, while weak ion exchangers are ionized within a narrow pH range. Thus, weak exchangers offer more flexibility in selectivity than do strong ion exchangers, although the strong ion exchangers are used for initial development and optimization, because binding capacity does not change with pH. For the separation, the column is equilibrated with a start buffer, and then an analyte containing an opposite charge binds to the ionic groups of the matrix, whereas uncharged molecules, or those with the same charge as the ionic groups, are not retained. The adsorbed analyte of interest can be eluted by a gradient of ionic strength, pH values, or a combination of both in the mobile phase. The action mechanism of ion exchanger is shown in Figure 5 (Oliver, 1991; GE Healthcare, 2004; Meyer, 2010).

**Size-exclusion chromatography (SEC)** is a preparative and non-destructive analytical technique that, unlike other methods, is not based on interactions between molecules and the stationary phase, but on the size of molecules (Figure 6). The column is packed with inert material with pores of controlled size, within the stationary phase, such that the small molecules can enter most of the pores and therefore will be retained the longest time, while the larger molecules cannot penetrate and are kept for a shorter time period. The SEC can be classified according to the mobile phase used in *gel filtration chromatography* or *gel permeation chromatography*. Gel filtration chromatography (GFC) uses an aqueous mobile phase, which may contain organic modifiers or salts to change the ionic force or buffer solutions to change pH. Gel permeation chromatography (GPC) is a method used to separate high polymers, and it has become a prominent and widely used method for estimating molecular-weight distributions. Unlike GFC, GPC uses organic mobile phases such as tetrahydrofuran (THF), toluene, chloroform, dichloromethane, or dimethylformamide (Oliver, 1991; Meyer, 2004; GE Healthcare, 2010).

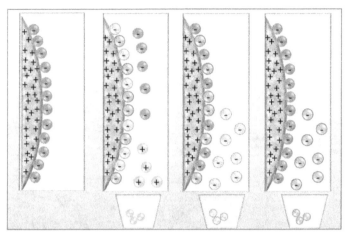

Fig. 5. The principle of IEC Separation. The mobile phase has ions negatively charged (red circles) that binding to the stationary phase (positively charged). The sample containing mixture of positively and negatively charged groups flows through the column. Those analytes containing negative charge are able to displace the mobile phase ions and bind to the stationary phase, while the positives groups (yellow) are eluted. The sample bound in the stationary phase can be eluted by increasing the concentration of a similarly charged species. The analyte that binds weakly (green) in the stationary phase will be eluted by buffer with salt ions at lower concentration. The analyte that binds strongly (blue) in the stationary phase will be eluted by buffer with salt ions at higher concentration.

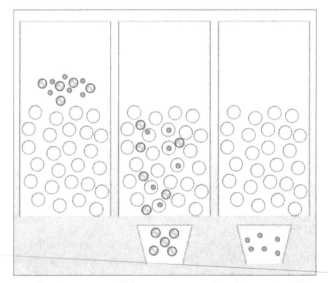

Fig. 6. Molecules smaller than pore will become trapped in the matrix. Those of larger molecular weight will not be trapped but will flow through column. Thus, larger molecules elute first, while smaller molecules are held longer inside the pores and will be eluted last.

## 3.1 Liquid chromatography coupled to MS

In proteomics, LC can be performed downstream of 2D gels to fractionate peptides from excised spots before MS analysis, upstream of 2D gels to prefractionate the protein sample, or instead of 2D gels as the main separation method in a multidimensional protein identification technology (MudPIT) experiment (item 4). Two strategies for protein identification and characterization by MS currently are employed in proteomics. In the bottom-up strategy, purified proteins or complex protein mixtures are subjected to enzymatic digestion, and the peptide products are analyzed by MS (Andersen et al., 1996; Pandley & Man, 2000) (Figure 7). In top-down proteomics, intact proteins or big protein fragments are subjected to fragmentation during MS analysis (Kelleher, 2004; Siuti & Kelleher, 2007). The major problem in bottom-up proteomics is that many peptides are generated for subsequent mass analysis, so it is not possible to get full protein sequence coverage and protein inference can be a problem. Different proteins can share some peptides, and consequently identification through bottom-up proteomics could be ambiguous. For these reasons, many efforts have focused on separation methods such as multidimensional chromatography to improve sensitivity and resolution. On the other hand, top-down proteomics permits high coverage sequence, thus overcoming one of the most important challenges in bottom-up strategy; however, it is a newer approach with several instrument limitations that will benefit from some hardware MS advances.

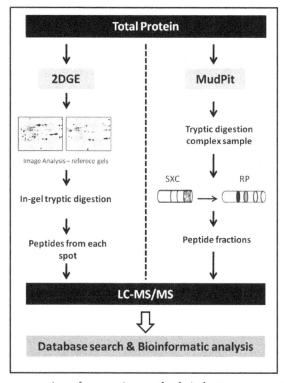

Fig. 7. Schematic representation of separation methods in bottom-up proteomics approaches (Adapted from Agrawal et al.,2008).

Also, the so-called "shotgun proteomics" is applied in a typical bottom-up approach and involves the utilization of HPLC coupled to tandem mass spectrometry (MS/MS) for identification of proteins on a large scale. This platform has been facilitated by the use of MudPIT, discussed below. As bottom-up proteomics is the most mature and most widely used approach for protein identification, in the following section we focus on chromatographic platforms for peptide separation.

## 4. Multidimensional liquid chromatography in proteomics

As previously mentioned, bottom-up proteomics offers some challenges to be overcome, such as sample complexity and large differences in protein concentration. As we know, these problems are far from being solved by single chromatographic or electrophoretic methods. Basically, two main approaches have been developed to face these difficulties: methods to separate abundant proteins from lower abundance proteins and multidimensional separation methods to maximize fractionation of peptides, thus increasing the proteome coverage analyzed by MS.

Abundant proteins can be a big problem when analyzing complex samples from certain tissues. Some tissues have high percentage (around 50% or more) of some classes of proteins. For example, leaves show approximately 50% of Rubisco (Feller et al., 2008) and the great majority of human blood serum proteome is comprised of albumin, fibrinogen, and immunoglobulins, transferrin, haptoglobin, and lipoproteins (Burtis & Ashwood, 2001; Turner & Hulme, 1970). If these abundant proteins are not removed from the sample, the peptides resulting from these proteins for proteomics analysis will overlap the peptides from the low-abundance proteins. To remove abundant proteins in complex samples, immunoaffinity separation or gel-based fractionation could be good options, although immunoaffinity separation can also remove low- abundance proteins that have any level of interaction with those abundant proteins (Granger et al., 2005).

Sequential chromatographic separations utilizing different chemical or physical properties have improved the assessment of classes of proteins that are difficult to handle in gel electrophoresis. MDLC can reach the same resolution of 2D gel electrophoresis with added advantages like automation, better sensitivity, and increased proteome coverage. MDLC was first described by Giddings in 1984 (Giddings, 1984) as a technique which combines two or more types of LC to increase the peak capacity and selectivity, contributing to a better fractionation of peptides that will enter the mass spectrometer.

The peak capacity is the number of peaks that can be resolved in a given time. To increase the peak capacity, the combination of two or more orthogonal separation dimensions is necessary. In other words, the properties affecting the separation in the first dimension do not affect the separation in the subsequent dimensions; thus, the process simplifies the sample complexity and improves the power of resolution, fractionating more components in a given time.

Many LC combinations have been reported utilizing chromatographic methods like strong cation exchange (SCX), strong anion exchange (SAX), size-exclusion (SEC) and reverse-phase chromatography (Zhang et al., 2007; Hynek et al., 2006; Moritz et al., 2005). Some factors are essential when considering an MDLC approach. The choice of different types of chromatographic columns to reach a satisfactory MDLC peak capacity is the most important point within this strategy.  In order to separate a large variety of peptides in high-performance chromatography, the columns used in each dimension have to work with no correlated properties. Other factors to be considered include the fact that the first column should have a larger loading capacity and have solvent compatibility with the subsequent columns. Thus, this compatibility between dimensions is essential in online systems. The last step of chromatography immediately before MS analysis should be compatible with electrospray ionization (ESI). For this purpose, the reverse-phase (RP) chromatography is frequently used, as it can desalt the samples and it is completely compatible with ESI, thus providing effective resolution and facilitating MS detection.

Several proteomic studies using different multidimensional configurations have been done along these years (reviewed by Gao et al., 2010) Here we point out some examples of the most applicable approaches.

## 4.1 Ion-exchange and reverse-phase chromatography (IEX-RPLC)

MUDPIT was first introduced by Yates and co-workers (Link et al., 1999; Wolters et al., 2001) using a biphasic column sequentially packed with SCX particles and then with C18 particles for peptide separation prior to MS analysis. Other works reported the use of an SCX column and an RP column connected to perform an online SCX-RP-MS/MS analysis (Tram et al., 2008; Gilar et al., 2005). Besides their orthogonality, SCX and RP columns show mobile phase compatibility between each other and with MS analysis.

In the SCX-RP configuration, peptides from a complex mixture are acidified and applied to an SCX column, in which the elution steps can be done using a salt gradient. Then, a fraction of peptides are absorbed in the RP column, and after washing away salts and buffers, peptides are eluted from the RP column into the mass spectrometer using a gradient of an organic solvent. Finally, the RP is reequilibrated and new fraction of peptides from the SCX column is eluted to be absorbed in the RP column, and the process repeats.

Several studies came out to improve the application of the MudPIT technique. One of the limitations of SCX-RPLC is related to the use of salt gradient SCX separation, resulting in limited resolution and peak capacity for peptide separation. Extensive salt usage could also cause ion suppression, thus reducing MS performance. Second, by using coupled SCX-RP columns, separation on both dimensions has to be performed at the same flow rate, which may sacrifice MS detection sensitivity for low-abundance components. As an alternative approach, offline SCX-RP separation has been used to identify more than 1200 proteins from zebrafish liver (Wang et al., 2007).However, the implementation of offline configurations is not always the best option, as extensive offline sample handling increases the overall analysis time and causes sample loss and sensitivity reduction.

Dai et al. (2005) reported an integrated column composed of SCX and RP where peptides were fractionated by a pH step gradient. The exclusion of salt removal steps could lead to fast 2D-LC separation and MS/MS detection. Some years later, the same research group developed a SCX and SAX combined pre-fraction strategy called Yin-Yang multidimensional chromatography (Dai et al., 2007). Peptides eluted from a SCX column at pH 2.5 were then separated by a SAX column using a pH gradient solution. Subsequently, SAX fractions were analyzed by RPLC-MS/MS. This approach revealed proteins from a broad range of different pH values.

## 4.2 Reverse-phase-reverse-phase chromatography (RP-RPLC)

The use of RP columns in both dimensions of a MudPIT analysis exhibited high-throughput, automatability, and performance comparable with that of SCX-RP. Despite all successful studies based on SCX-RP, these approaches are frequently encountering higher complexity samples than can be accommodated by their power of resolution. Accordingly, new strategies have been developed, including the RP-RPLC. Gilar et al. (2005) investigated the use of a pH gradient in a RP-RP platform. The RP-RP approach provides higher peak capacity in both dimensions, and the pH gradient has the most significant impact on the selectivity of this platform. The pH gradient modulates the peptide hydrophobicity during the elution of fractions of peptides.

Another factor that has been increasing the sensitivity of this approach is the integration of a high-pressure system in RP-RP platforms. This was first implemented in the Agilent 1100 2-D liquid chromatographic system and more recently by the 2D NanoACQUITTY system from Waters, frequently used with a pH gradient and two columns of C18 particles. Zhou et al. (2010) revealed that RP-RP fractionation outperforms SXC-RP. It was also demonstrated that the combination of RP-RP systems with the nanoflow format had good impact in the efficiency of electrospray ionization prior to MS/MS analysis.

## 4.3 Affinity chromatography

Another type of chromatography largely used in proteomics is affinity-based. The main usage of this technique is to enrich post-translationally modified (PTM) peptides or proteins subject to MS analysis. Due to their low stoichiometry and dynamic modification patterns, PTM materials have to be enriched in complex mixtures prior to MS analysis. Affinity chromatographic columns can be used online or offline to RP columns and the mass spectrometer.

## 4.4 Enrichment of phosphopeptides or phosphoproteins

In phosphoproteomics, the interest focuses on the identification and quantification of phosphorylated proteins or peptides. With this purpose, several selective enrichment methods are applied to increase the content of phosphopeptides or phosphoproteins in complex mixtures, thus preventing the suppression of ion signals by unphosphorylated molecules.

Immobilized metal affinity chromatography (IMAC) is one of the most popular techniques employed in phosphoproteomics. IMAC contains immobilized positive ions ($Fe^{3+}$, $Ga^{3+}$ and $Al^{3+}$) interacting with the negative ions of phosphate groups. The elution of the interacted molecules makes the enrichment possible. Tsai et al. (2011) showed a pH/acid-controlled IMAC protocol for phosphopeptide purification with high specificity and lower sample loss. They characterized over 2,360 nondegenerate phosphopeptides and 2,747 phosphorylation sites in the H1299 lung cancer cell line, showing that a low pH buffer increases the specificity of IMAC for phosphopeptides. The buffer composition was directly associated to the specificity and selectivity of the IMAC technique (Tsai et al., 2008; Jensen et al., 2007).

Another important enrichment technique widely used is the metal oxide affinity chromatography (MOAC). MOAC uses a principle similar to that of IMAC, incorporating metal oxides, such as titanium dioxide $TiO_2$, zirconium dioxide $ZrO_2$, or aluminum hydroxide $Al(OH)_3$. Metal oxides tend to have higher selectivity for phosphopeptides, making it easier to trap them in the column (Zhou et al., 2007). During the last few years, $TiO_2$ has emerged as the main MOAC-based phosphopeptide enrichment method (Pinkse et al., 2004). The principle is the same as that of IMAC; however, when loading peptides with DHB (2,5-dihydroxybenzoic acid) non-specific binding is reduced, thus increasing the selectivity of $TiO_2$ (Larsen et al., 2005).

## 4.5 Enrichment of glycopeptides or glycoproteins

Glycosylation is the most complex PTM presented in eucaryotes. Like other PTMs, glycopeptides are often very minor constituents compared to peptides derived from proteolytic digestion. Therefore, enrichment of glycoproteins or glycopeptides is essential in a glycoproteomics study (Ito et al., 2009). The lectin affinity approach is the most common tool for glycoprotein/glycopeptide enrichment. Lectins are sugar-binding proteins that are highly specific to sugar moieties. For example, concanavalin A (conA) is a mannose-binding lectin largely used to study N-glycosylated proteins, while lectin from *Vicia villosa* (VVL), which preferentially binds to alpha- or beta-linked terminal GalNAc, has a different preference. All lectins have sugar specificity, and therefore serial lectin affinity columns (SLAC) have been developed to reduce the complexity of proteolytic digests by more than one order of magnitude. SLAC can be used to study O-glycosylation proteins, which are difficult to access because lectins for studying them are not specific enough. Jacalin, a lectin, is relatively specific for O-glycosylation but has the problem that it also selects mannose N-glycans. This problem can be overcome by first using a ConA affinity column to first remove mannose, and then using Jacalin columns. When used in a serial configuration, O-glycosylated peptides can be accessed (Durham & Regnier, 2006).

## 4.6 LC-MS repeatability and reproducibility

First of all, it is important to understand the difference between repeatability and reproducibility. The first, represents the variations in the measurements on the same sample, made in the same instrument and by the same operator (Bland & Altman, 1986). The second, is the variation observed for an analytical technique when operator, instrumentation, time, or location is changed (McNaught & Wilkinson, 1997). So in proteomics, it is possible to calculate variations in results from run to run estimating the repeatability of the analytical technique or estimating the reproducibility between different laboratories in completely differently instruments. These measurements are imperative to the clinical utility of biomarker candidates and must be expected (Baggerly et al.,2005).

During proteomics analysis there are many potential contributors to variability that can compromise the approach repeatability and reproducibility. The variations can begin in sample collection, specimen processing techniques, storage, and instrument performance (Hsieh at al, 2006; Banks et al., 2005; Pilny et al., 2006). All steps in the proteome analysis can offer a source of variations. The proteomics analysis by LC-MS begins with the digestion of complex mixtures followed by peptide fractionation in LC systems. Then MS/MS scans are acquired and spectra matches are resulted from bioinformatics analysis. The complexity of these steps leads to variations in peptide and protein identification and quantification. Minor differences in LC can result in modified peptide elution times (Prakash et al., 2006) or change which peptides are selected for MS/MS fragmentation (Liu et al., 2004).

In this context, the use of MudPIT also can introduce more variation. Delmotte et al. (2009) showed that an introduction of a separation dimension decreases the repeatability by approximately 25% upon 1D or 2D chromatographic separations. Slebos et al. (2008) reported superior reproducibility in isoelectric focusing (IEF) compared to SCX separations. They found that IEF more quickly reached maximal detection within three replicate analysis. In contrast, the SCX required six replicates. In this study, approximately 90% of all peptide identifications are found in a single fraction. In contrast, SCX is characterized by spread of peptides into adjacent fractions. Peptides at lower abundance or those generating lower signal intensity are more likely to be selected for MS/MS if they appear in multiple fractions. Thus SCX produces more chances of peptides detection which is consequence of the greater SXC sensitivity of peptide identifications. Though, greater sensitivity leads to much higher variability in peptide detection at each fractionation step from run to run, demanding a greater number of technical replicas, which compromises the overall throughput. Several groups have evaluated the number of replicates necessary to observe a particular percentage of the proteins in a sample (Liu et al., 2004; Slebos et al., 2008; Kislinger et al., 2005).

On the other hand, Tabb et al. (2009) concluded that a standardized platforms results a high degree of repeatability and reproducibility (around 70-80%) in protein identifications. This is an indication that LC-MS/MS platforms can generate consistent protein identifications. They showed that instrumentation can also increment variations. In this study, they observed that the high resolution of Orbitraps outperforms lower resolution of LTQs in repeatability and reproducibility. They also observed that reproducibility between different instruments of the same type is lower than repeatability of technical replicates on a single instrument by several percent.

Standardizations of methodologies and system configurations, as well as appropriate instrument tuning and maintenance would result in lower noise level from biological, chemical or instrumental sources between LC-MS/MS analysis. A comprehensive understanding about components that affect reproducibility and repeatability can help in the determination of the best alternative of experimental design to reach the reliability desired in proteomics studies.

## 5. Concluding remarks

Protein separation methods are vital for the characterization of proteomes. All these methods exploit one or more general properties of proteins and are directly linked to the effectiveness of any proteomic analysis. Traditionally, 2D gels have been the most frequently employed protein separation tool, capable of separating up to 10,000 components. However, 2D gels have some limitations, as discussed above, preventing efficient automation. In addition, LC methods coupled to MS analysis have been emerged as a promising strategy to achieve better automation and sensitivity. Combining orthogonal chromatographic columns increases peak capacity and permits a large dynamic range to be more efficiently measured in complex samples. Also, affinity chromatography allows the selection of a subgroup of proteins (proteins with certain PTMs), directing the study into specific biological pathways and increasing the generation of relevant information. Although its resolution has been improved resolution, LC fractioning has several concerns to be circumvented, such as the amount of time required, reproducibility, and protein inference from peptides. An experiment involving RP analysis can take 1.5 to 2 hours for a single analysis. Then, considering 15 fractions (taken from an IEX) from a complex mixture, there will be 22.5 hours required. Moreover, there is the analysis of biological (10 individual) and technical replicas (3 injections) that could reach 56 days of analysis, taking into account an experiment comparing control and affected. The required time increases instrument and maintenance costs and influences the experimental throughput. The second concern involves reproducibility during fractionating, which can be related to the number of dimensions employed and if they are inline coupled to the mass spectrometer or not. Another point that has received considerable attention is the development of data analysis pipelines to effectively process LC data from multiple peptide fractionation. The challenge is to reach better accuracy in protein inference, mainly in bottom-up proteomics. In parallel, top-down proteomics is developing rapidly, with significant progress in MDLC systems for protein separation and probably warranting more attention in the near future.

Finally, the achievement of better representation, resolution, sensitivity, and automation has been developed with the application of LC methods coupled to MS, but they are far from being considered as the perfect strategy and are subject to improvements. The use of 2D gels or LC as the central separation method or both in combination is debatable, depending on the study objective. However, the constant advancement of these platforms is resulting in successful proteomics studies.

# 6. References

Adam, B.L.; Qu, Y.; Davis, J.W.; Ward, M.D.; Clements, M.A.; Cazares, L.H.; Semmes, O.J.; Schellhammer, P.F.; Yasui, Y.; Feng, Z. & Wright, G.L.Jr. (2002). Serum protein fingerprinting coupled with a pattern-matching algorithm distinguishes prostate cancer from benign prostate hyperplasia and healthy men. *Cancer Research*, Vol.62, No.13, (July 2002), pp.3609–3614, ISSN 1541-7786

Agrawal, G.K.; Hajduch, M.;, Graham, K.; Thelen, J.J.(2008) In-Depth Investigation of the Soybean Seed-Filling Proteome and Comparison with a Parallel Study of Rapeseed. *Plant Physiology*, Vol. 148, pp. 504–518

Albrethsen, J.; Bogebo, R.; Olsen, J.; Raskov, H.; Gammeltoft, S. (2006). Preanalytical and analytical variation of surface-enhanced laser desorption-ionization time-of-flight mass spectrometry of human serum. *Clinical Chemistry and Laboratory Medicine*, Vol.44, No.10, pp. 1243-52

Andersen, J. S., Svensson, B., and Roepstorff, P. (1996) Electrospray ionization and matrix assisted laser desorption/ionization mass spectrometry: Powerful analytical tools in recombinant protein chemistry. *Nature Biotechnology.* Vol.14, pp. 449–457

Anderson, N.L.; Polanski, M.; Pieper, R.; Gatlin, T.; Tirumalai, R.S.; Conrads, T.P.; Veenstra, T.D.; Adkins, J.N.; Pounds, J.G.; Fagan, R. & Lobley, A. (2004). The Human Plasma Proteome: A Non-Redundant List Developed by Combination of Four Separate Sources, *Mol. Cell Proteomics*, Vol. 3, pp. 311–326

Baggerly, K. A.; Morris, J. S.; Edmonson, S. R.; Coombes, K. R. (2005) Signal in noise: evaluating reported reproducibility of serum proteomic tests for ovarian cancer. *Journal of the National Cancer Institute*, Vol.97, No.4, pp. 307-9

Banks, R. E.; Stanley, A. J.; Cairns, D. A.; Barrett, J. H.; Clarke, P.; Thompson, D.; Selby, P. J. (2005). Influences of blood sample processing on low-molecular-weight proteome identified by surface-enhanced laser desorption/ionization mass spectrometry. *Clinical Chemistry*, Vol. 51, No.9, pp. 1637-49

Bianco, C.; Strizzi, L.; Mancino, M.; Rehman, A.; Hamada, S.; Watanabe, K.; De Luca, A.; Jones, B.; Balogh, G.; Russo, J.; Mailo, D.; Palaia, R.; D'Aiuto, G.; Botti, G.; Perrone, F.; Salomon, D.S. & Normanno, N. (2006). Identification of cripto-1 as a novel serologic marker for breast and colon cancer. *Clinical Cancer Research*, Vol.12, No.17, (September 2006), pp.5158-64, ISSN 1557 -3265

Bjellqvist, B.; Ek, K.; Righetti, P.G.; Gianazza, E.; Görg, A.; Westermeier, R. & Postel, W. (1982).Isoelectric-focusing in immobilized ph gradients - principle, methodology and some applications. *Journal of Biochemical and Biophysical Methods*, Vol.6, No.4, (September 1982), pp.317-339 ISSN 0165-022X

Bland, J. M.; Altman, D. G. (1986). Statistical methods for assessing agreement between two methods of clinical measurement. *The Lancet*, Vol. 1, No. 8476, pp. 307–310

Burtis,C.A. & Ashwood E. R. (Eds.). (2001). *Tietz Fundamentals of Clinical Chemistry*, 5th Edition, WB Saunders, ISBN 0-7216-8634-6, Philadelphia

Chevalier, F. (2010). Highlights on the capacities of "Gel-based" Proteomics. In: *Proteome Science*, Vol.8, No.23, 29/07/2011, Available from: <http://www.proteomesci.com/content/8/1/23>

Dai, J.; Jin, W.H.; Sheng, Q.H.; Shieh, C.H.; Wu, J.R. & Zeng, R. (2007). Protein phosphorylation and expression profiling by Yin-yang multidimensional liquid chromatography (Yin-yang MDLC) mass spectrometry. *Journal of Proteome Research*, Vol.6, pp. 250-62

Dai, J.; Shieh, C.H.; Sheng, Q.; Zhou, H. & Zeng, R. (2005). Proteomic analysis with integrated multiple dimensional liquid chromatography/mass spectrometry based on elution of ion exchange column using pH steps, *Analytical Chemistry*, Vol.77, pp. 5793-5799

Delmotte, N.; Lasaosa, M.; Tholey, A.; Heinzle, E.; van Dorsselaer, A.; Huber, C. G. (2009). Repeatability of peptide identifications in shotgun proteome analysis employing off-line two-dimensional chromatographic separations and ion-trap MS, *Journal of Separation Science.*, Vol. 32, No.8, pp. 1156–11164

Drews, O.; Reil, G.; Parlar, H. & Görg, A. Setting up standards and a reference map for the alkaline proteome of the Gram-positive bacterium *Lactococcus lactis*. (2004). *Proteomics*, Vol.4, No.5, (May 2004), pp.1293-1304, ISSN 1615-9853

Durham, M. & Regnier F. E. (2006). Targeted glycoproteomics: serial lectin affinity chromatography in the selection of O-glycosylation sites on proteins from the human blood proteome. *Journal of chromatography A*, Vol. 1132, pp. 165-73.

Elia, G. (2008). Biotinylation reagents for the study of cell surfasse Proteins. *Proteomics*, Vol. 8, pp. 4012–4024

Feller, U.; Anders, I. & Mae, T. (2008). Rubiscolytics: fate of Rubisco after its enzymatic function in a cell is terminated. *Journal of Experimental Botany*, Vol.59, pp. 1615–1624

Figeys, D. (2005). Proteomics: The Basic Overview. In: *Industrial Proteomics: Applications for Biotechnology and Pharmaceuticals*, edited by Daniel Figeys, pp.1-62, John Wiley & Sons Incorporated, ISBN 0471703753, San Francisco, USA

Gao, M.; Qi, D.; Zhang, P.; Deng, C. & Zhang, X. (2010). Development of multidimensional liquid chromatography and application in proteomic analysis, *Expert Review of Proteomics*, Vol.7, No.5, pp.665-678

GE Healthcare. (2004). *Ion Exchange Chromatography & Chromatofocusing*, Amersham Biosciences Limited, 11-0004-21, Edition AA, Sweden.

GE Healthcare. (2006). *Hydrophobic Interaction and Reversed Phase Chromatography, Principles and Methods*, GE imagination at work, 11-0012-69 AA, Sweden.

GE Healthcare. (2007). *Affinity Chromatography, Principles & Methods*, GE imagination at work, 18-1022-29 AE, Sweden.

GE Healthcare. (2010). *Gel Filtration, Principles and Methods*, GE imagination at work, 18-1022-18 AK, Sweden.

Giddings, J.C. (1984). Two-Dimensional Separations: Concept and Promise, *Anal. Chem.*, Vol.56, pp.1258A–1270A

Gilar, M.; Olivova; P., Daly; A. E. & Gebler, J. C. (2005). Two-dimensional separation of peptides using RP-RP-HPLC system with different pH in first and second separation dimensions, *Journal of Separation Science*, Vol.*28*, pp. 1694–1703

Görg, A.; Obermaier, C.; Boguth, G.; Csordas, A.; Diaz, J.J. & Madjar, J.J. Very alkaline immobilized pH gradients for two-dimensional electrophoresis of ribosomal and nuclear proteins. (1997). *Electrophoresis*, Vol.18, No.3-4, (March-April 1997), pp.328-337, ISSN 1522-2683

Görg, A.; Obermaier, K.; Boguth,G.; Harder, A.; Scheibe, B.; Wildgruder, R. & Weiss, W.(2000). The current state of two-dimentional electrophoresis with immobilized pH gradients. *Electrophoresis*, Vol.21, No.6, (April 2000), pp.1037-1053, ISSN 1522-2683

Granger, J.; Siddiqui, S.; Copeland, S. & Remick D. (2005). Albumin depletion of human plasma also removes low abundance proteins including the cytokines *Proteomics*, Vol.5, pp. 4713–4718.

Greenough, C.; Jenkins, R.E.; Kitteringham, N.R.; Pirmohamed, M.; Park, B.K. & Pennington, S.R. (2004). A method for the rapid depletion of albumin and immunoglobulin from human plasma. *Proteomics*, Vol.4, No.10, (October 2004), pp.3107-3111, ISSN 1615-9853

Guo, X.; Zhao, C.; Wang, F.; Zhu, Y.; Cui, Y.; Zhou, Z.; Huo, R. & Sha, J. (2010). Investigation of human testis protein heterogeneity using 2-dimensional electrophoresis. *Journal of Andrology*, Vol31, No.(4), pp.419-429, (July-August 2010), ISSN 0196-3635

Hsieh, S. Y.; Chen, R. K.; Pan, Y. H.; Lee, H. L. (2006). Systematical evaluation of the effects of sample collection procedures on low-molecular-weight serum/plasma proteome profiling. *Proteomics*, Vol. 6, No.10, pp.3189-98

Hynek, R.; Svensson, B.; Jensen, O.N.; Barkholt, V. & Finnie, C. (2006). nrichment and identification of integral membrane proteins from barley aleurone layers by reversed-phase chromatography, SDS-PAGE and LC-MS/MS, *Journal Proteome Res.*, Vol.5, pp. 3105–3113

Ito, S.; Hayama, K. & Hirabayashi, J. (2009). Enrichment strategies for glycopeptides, *Methods in Molecular Biology,*Vol.534, pp.195-203

Jang, J.H. & Hanash, S. (2003). Profiling of the cell surface proteome. *Proteomics*, Vol.3, No.10, (October 2003), pp.1947-1954, ISSN 1615-9853

JBC Centennial 1905–2005 100 Years of Biochemistry and Molecular Biology. (2006). The Development of Two-dimensional Electrophoresis by Patrick H. O'Farrell. *The Journal of Biological Chemistry*, Vol.281, No.32, (August 2006), pp. e26, ISSN 1083-351X

Jensen S.S. & Larsen M.R. (2007). Evaluation of the impact of some experimental procedures on different phosphopeptide enrichment techniques. *Rapid Commun. Mass Spectrom,* Vol. 21, pp. 3635–45

Klein-Scory, S.; Kübler, S.; Diehl, H.; Eilert-Micus, C.; Reinacher-Schick, A.; Stühler, K.; Warscheid, B.; Meyer, H.E.; Schmiegel, W. & Schwarte-Waldhoff, I. (2010). Immunoscreening of the extracellular proteome of colorectal cancer cells. *BMC Cancer*, Vol.10, No.70, pp.2-19, (February 2010), ISSN 1471-2407

Larsen M.R.; Thingholm T.E.; Jensen O.N.; Roepstorff P. & Jorgensen TJ. (2005). Highly selective enrichment of phosphorylated peptides from peptide mixtures using titanium dioxide microcolumns. *Mol. Cell Proteomics,* Vol. 4, pp. 873–86

Lilley, K.S. & Friedman, D.B. (2004). All about DIGE: quantification technology for differential-display 2D-gel proteomics. *Expert Review of Proteomics*, Vol.1, No.4, (December 2004), pp.401-409, ISSN 1478-9450

Link, A.J.; Eng, J.; Schieltz, D.M.; Carmack, E.; Mize, G.J.; Morris, D.R.; Garvik, B.M. & Yates, J.R. III (1999). Direct analysis of protein complexes using mass spectrometry, *Nature Biotechnology*, Vol.17, pp. 676–682

Liu, H.; Sadygov, R. G.; Yates, J. R., III. (2004). A model for random sampling and estimation of relative protein abundance in shotgun proteomics. *Analytical Chemistry.*, Vol.76, No.14, pp.4193–4201

Kelleher, N. L. (2004). Top-Down Proteomics, *Analytical Chemistry*, Vol.76, No.11, pp.197A – 203A

Kislinger, T.; Gramolini, A. O.; MacLennan, D. H.; Emili, A. (2005). Multidimensional protein identification technology (MudPIT): technical overview of a profiling method optimized for the comprehensive proteomic investigation of normal and diseased heart tissue. *Journal of the American Society for Mass Spectrometry*, Vol. 16, No. (8), pp. 1207–1220

McNaught, A. D.; Wilkinson, A. (1997). *Compendium of Chemical Terminology (the "Gold Book")*., 2.02 ed.; Blackwell Scientific Publications: Oxford, ISBN 0-86542-684-8, North Carolina, United States of America

Meyer,V.R. (2010). *Practical High-Performance Liquid Chromatography (5th)*, Wiley, ISBN 978-0-470-68218-0, United Kingdom

Molloy, M.P. (2000). Two-dimensional electrophoresis of membrane proteins using immobilized pH gradients. *Analytical Biochemistry*, Vol.280, No.1, (April 2000), pp.1-10, ISSN 1096-03

Moritz, R.L. ; Clippingdale, A.B.; Kapp, E.A.; Eddes, J.S.; Ji, H.; Gilbert, S.; Connolly, L.M. & Simpson, R.J. (2005). Application of 2-D free-fl ow electrophoresis/ RP-HPLC for proteomic analysis of human plasma depleted of multi-high abundance proteins, *Proteomics*, Vol.5, pp. 3402–3413

Neue, U. D. (1997). *HPLC Columns, Theory, Technology and Practice*, Wiley, ISBN 978-0-471-19037-0, New York

O'Farrell, P. (1975). High Resolution Two-Dimensional Electrophoresis of Proteins. *The Journal of Biological Chemistry*, Vol.250, No.10, (May 1975), pp.4007–4021, ISSN 1083-351X

Oliver, R.W.A. (1991). *HPLC of macromolecules: a practical approach*, IRL press, ISBN 0199630208, England.

Pandey, A., and Mann, M. (2000) Proteomics to study genes and genomes. *Nature*, Vol.405, pp.837–846

Patton, W.F. (2002) Detection technologies in proteome analysis. *Journal of Chromatography B-Analytical Technologies in the Biomedical and Life Sciences*, Vol.771, No.(1-2), (May 2002), pp.3-31, ISSN 1873-376X

Pilny, R.; Bouchal, P.; Borilova, S.; Ceskova, P.; Zaloudik, J.; Vyzula, R.; Vojtesek, B.; Valik, D. (2006). Surface-enhanced laser desorption ionization/time-of-flight mass spectrometry reveals significant artifacts in serum obtained from clot activator-containing collection devices. *Clinical Chemistry*, Vol.52,No.11, pp.2115-6

Pinkse M.W.; Uitto P.M.; Hilhorst M.J.; Ooms B. & Heck A.J. (2004). Selective isolation at the femtomole level of phosphopeptides from proteolytic digests using 2D-NanoLC-ESI-MS/MS and titanium oxide precolumns. *Anal. Chem.* Vol.76, pp. 3935–43

Rabilloud, T.; Chevallet, M.; Luche, S. & Lelong, C. (2008). Fully denaturing two-dimensional electrophoresis of membrane proteins: A critical update. *Proteomics,* Vol. 8, pp. 3965–3973

Prakash, A.; Mallick, P.; Whiteaker, J.; Zhang, H.; Paulovich, A.; Flory, M.; Lee, H.; Aebersold, R.; Schwikowski, B. (2006). Signal maps for mass spectrometry-based comparative proteomics, *Mol. Cell.Proteomics,* Vol.5, No.3, pp.423–432

Rogers, S.; Girolami, M.; Kolch, W.; Waters, K.M.; Liu, T.; Thrall, B. & Wiley, H.S. (2008). Investigating the correspondence between transcriptomic and proteomic expression profiles using coupled cluster models *Bioinformatics,* Vol.24, pp. 2894–2900

Saito, Y.; Jinno, K.; Greibrokk, T. (2004). Capillary columns in liquid chromatography: between conventional columns and microchips, *Journal of Separation Science,* Vol. 27, No.17-18, (December 2004), pp. 1379-1390, ISSN 1615-9306.

Santoni, V.; Molloy, M. & Rabilloud, T. (2000). Membrane proteins and proteomics: Un amour impossible? *Electrophoresis,* Vol.21, No.6 , (April 2000), pp.1054-1070, ISSN 1522-2683

Shaw, J.; Rowlinson, R.; Nickson, J.; Stone, T.; Sweet, A.; Williams, K. & Tonge, R. (2003). Evaluation of saturation labelling two-dimensional difference gel electrophoresis fluorescent dyes. *Proteomics,* Vol.3, No.7, (July 2003), pp.1181-1195, ISSN 1615-9853

Siuti, N; Kelleher, N .L. (2007). Decoding protein modifications using top-down mass spectrometry, *Nature Methods,* Vol.4, pp.817-821

Slebos, R. J.; Brock, J. W.; Winters, N. F.; Stuart, S. R.; Martinez, M. A.; Li, M.; Chambers, M. C.; Zimmerman, L. J.; Ham, A. J.; Tabb, D. L.; Liebler, D. C. (2008). Evaluation of strong cation exchange versus isoelectric focusing of peptides for multidimensional liquid chromatography-tandem mass spectrometry. *J. Proteome Res.,* Vol.7, No.12, pp.5286–5294

Skoog, D.A.; Holler, J.F.; Crouch S.R. (2006). Liquid Chromatography, In: *Principles of Instrumental Analysis (6th),* Skoog, D.A.; Holler, J.F.; Crouch, Thomson Brooks/Cole (David Harris), pp. (816-855), ISBN 9780495012016, Canada.

Smejkal, G.B. (2004) The Coomassie chronicles: past, present and future perspectives in polyacrylamide gel staining. *Expert Review of Proteomics,* Vol.1, No.4, (December 2004), pp.381-387, ISSN 1478-9450

Tan, S.; Tan, H.T. & Chung, M.C.M. (2008). Membrane proteins and membrane proteomics. *Proteomics,* Vol. 8, pp. 3924–3932

Tabb, D. L.; Veja-Montoto, L.; Rudnick, P.A.; Variyath, A. M. et al. (2010). Repeatability and Reproducibility in Proteomic Identifications by Liquid Chromatography-Tandem Mass Spectrometry, *Journal of Proteome Research,* Vol.9, pp. 761–776

Tran, B. Q.; Loftheim, H.; Reubsaet, L.; Lundanes, E. & Greibrokk, T. (2008). On-Line multitasking analytical proteomics: How to separate, reduce, alkylate and digest whole proteins in an on-Line multidimensional chromatography system coupled to MS, *Journal of Separation Science,* Vol.31, pp. 2913–2923

Tsai, C.F.; Wang, Y.T.; Chen, Y.R.; Lai, C.Y.; Lin, P.Y.; et al. (2008). Immobilized metal affinity chromatography revisited: pH/acid control toward high selectivity in phosphoproteomics. *Journal Proteome Res.*, Vol. 7, pp. 4058–69

Tsai, C.F.; Wang, Y.T.; Lin, P.Y. & Chen, Y.J. (2011). Phosphoproteomics by Highly Selective IMAC Protocol, *Neuroproteomics*, Series Neuromethods, Vol.57, pp.181-196, Pub. Date: May-25-2011

Turner, M. W. & Hulme, B. (1970) *The Plasma Proteins: An Introduction*, Pitman Medical & Scientific Publishing Co., Ltd., London

Ünlü, M.; Morgan, M.E. & Minden, J.S. (1997). Difference gel electrophoresis: a single gel method for detecting changes in protein extracts. *Electrophoresis*, Vol.18, No.11, pp.2071-2077, (October 1997), ISSN 1522-2683

Wang, N., MacKenzie, L., De Souza, A. G., Zhong, H. Y., Goss, G., Li, L.. (2007), Proteome profile of cytosolic component of zebrafish liver generated by LC-ESI MS/MS combined with trypsin digestion and microwave-assisted acid hydrolysis, *Journal of Proteome Research*, Vol.6, pp. 263–272

Wilkins, M.R.; Gasteiger, E.; Sanchez, J.C.; Bairoch, A. & Hochstrasser, D.F. (1998). Two-dimensional gel electrophoresis for proteome projects: the effects of protein hydrophobicity and copy number. *Electrophoresis*, Vol.19, No.8-9, (June 1998), pp.1501-1505, ISSN 1522-2683

Wolters, D.A.; Washburn, M.P. & Yates, J.R.III (2001). An automated multidimensional protein identification technology for shotgun proteomics, *Analytical Chemistry*, Vol.73, pp. 5683–5690

Yokoyama, R.; Iwafune, Y.; Kawasaki, H. & Hirano, H. Isoelectric focusing of high-molecular-weight protein complex under native conditions using agarose gel. (2009). *Analytical Biochemistry*, Vol.387, No.1, (April 2009), pp.60-63, ISSN 1096-0309

Zhang, J.; Xu, X.; Gao, M.; Yang, P. & Zhang, X. (2007). Comparison of 2-D LC and 3-D LC with post- and pre-tryptic-digestion SEC fractionation for proteome analysis of normal human liver tissue, *Proteomics*, Vol.7, pp. 500–512

Zhou, F.; Cardoza, J.D.; Ficarro, S.B.; Adelmant, G.O.; Lazaro, J.B.; Marto, J.A. (2010). Online Nanoflow RP-RP-MS Reveals Dynamics of Multicomponent Ku Complex in Response to DNA Damage, *Journal of Proteome Research*, Vol. 9, No.12, pp. 6242-55

Zhou, H. J.; Tian, R. J.; Ye, M. L.; Xu, S. Y.; Feng, S.; Pan, C. S.; Jiang, X. G.; Li, X. & Zou, H. F. (2007). Highly specific enrichment of phosphopeptides by zirconium dioxide nanoparticles for phosphoproteome analysis. *Electrophoresis*, Vol.28, No.13, pp. 2201-2215

# Part 2

# Sample Preparation

# 3

# A Critical Review of Trypsin Digestion for LC-MS Based Proteomics

Hanne Kolsrud Hustoft, Helle Malerod, Steven Ray Wilson,
Leon Reubsaet, Elsa Lundanes and Tyge Greibrokk
*University of Oslo*
*Norway*

## 1. Introduction

Proteomics is defined as the large-scale study of proteins in particular for their structures and functions (Anderson and Anderson 1998), and investigations of proteins have become very important since they are the main components of the physiological metabolic pathways in eukaryotic cells. Proteomics increasingly plays an important role in areas like protein interaction studies, biomarker discovery, cancer prevention, drug treatment and disease screening medical diagnostics (Capelo et al. 2009).

Proteomics can be performed either in a comprehensive or "shotgun" mode, where proteins are identified in complex mixtures, or as "targeted proteomics" where "selective reaction monitoring" (SRM) is used to choose in advance the proteins to observe, and then measuring them accurately, by optimizing the sample preparation as well as the LC-MS method in accordance to the specific proteins (Mitchell 2010).

Whether "MS-based shotgun proteomics" has accomplished anything at all regarding clinically useful results was recently addressed by Peter Mitchell in a feature article (Mitchell 2010), and he states that the field needs to make a further step or even change direction. Referring to discussions with among others John Yates and Matthias Mann, Mitchell addresses the failure in the search for biomarkers as indicators of disease, the difficulties of protein arrays, the uncertainty of quantification in "shotgun proteomics" (due to among others the efficiency of ionization in the mass spectrometers), database shortcomings, the problems of detecting post translational modifications (PTMs), and finally the huge disappointment in the area of drug discovery. The field points in the direction of targeted proteomics, but targeted proteomics will not be the solution to all our questions and comprehensive proteomics will still be needed. In order to get as much information, with as high quality as possible, from a biological sample, both the sample preparation and the final LC-MS analyses need to be optimized.

The most important step in the sample preparation for proteomics is the conversion of proteins to peptides and in most cases trypsin is used as enzyme. Trypsin is a protease that specifically cleaves the proteins creating peptides both in the preferred mass range for MS sequencing and with a basic residue at the carboxyl terminus of the peptide, producing information-rich, easily interpretable peptide fragmentation mass spectra. Some other proteases can be used as well, such as Lys-C, which is active in more harsh conditions with 8 M urea, and give larger fragments than trypsin. Asp-N and Glu-C are also highly sequence-specific proteases, but less active than the previously mentioned. Other less sequence-specific proteases are generally avoided since they create complex mixtures of peptides, difficult to interpret (Steen and Mann 2004). During a chromatographic separation of a complex mixture of peptides derived from a tryptic digestion, thousands of mass spectra are produced and sophisticated software is necessary to find matching proteins to the peptides identified. In complex proteomic samples, protein identification is performed by searching databases with search engines like Mascot, Sequest or Phenyx (IS 2011).

Protein identification traditionally follows two different workflows depending on the approach (Figure 1). In the gel electrophoresis-based approach the proteins are separated in one or two dimensions (1D/2D) on a gel and enzymatic digestion is performed in-gel, which is a time-consuming and tedious process (López-Ferrer et al. 2006). In the gel-free or in-solution based approach, the proteins or peptides, or both, are separated chromatographically using on-line LC systems and the proteins are digested in-solution

Fig. 1. Workflows of in-gel (left) and in-solution (right) digestion and subsequent LC-MS analysis of a protein sample.

(Capelo et al. 2009). The in-solution based approach tends to be the simplest in terms of sample handling and speed, but on the other hand it requires sophisticated LC-MS instrumentation which again requires constant maintenance.

The digestion step is the most time consuming step in the sample preparation workflow and different techniques to accelerate this procedure have been developed. Comparing these techniques, including some of our own experiments, the question of how we can evaluate the digestion efficiency materialized. The amino acid sequence coverage (SQ %) is often used as a measure for both the completeness of the protein digestion and the detection efficiency of the various tryptic peptides, and is a common way in proteomics to define the digestion rate (Xu et al. 2010). However, SQ % might be a misleading parameter to use, as different mass spectrometers and different search parameters in subsequent data analysis may reveal various SQ %. In addition it is of principal importance to relate SQ % to the degree of miss – cleavage peptides used to calculate this value: a high SQ % calculated from tryptic peptides without missed cleavages indicated a more complete digest than the same high SQ % calculated from tryptic peptides with many missed cleavages.

To get some information of the digestion efficiency, as a check before performing the data analysis, the possible presence of intact protein in the total ion chromatogram (TIC) may be used. However, this method can only be used for proteins small enough to be detected by the MS, such as cytochrome-C (cyt-C) (unless you have a MALDI MS available). On the other hand, evaluating the digestion rate this way, using an easily digested protein such as cyt-C, will give a good indication of the efficiency of the method; if an intact protein peak from cyt-C is detected in the chromatogram, then the digestion can be considered insufficient. Other non-protein reagents that are cleaved by trypsin might also be used as an internal standard when performing tryptic digestion of a complex sample, to have control over the digestion efficiency.

For quantitation of proteins it is necessary to find relevant indicators of their abundance in the mass spectrometer output. Several ways of protein quantitation have been suggested and they can be divided into two main categories; the isotope based and the label free methods. Two papers which give good overviews over the different labeling methods have been published recently (Capelo et al. 2010; Vaudel et al. 2010). In brief; the main modern strategies for isotopic labeling are divided into metabolic labeling at cell growth called SILAC, chemical labeling at protein level, called iCAT, enzymatic labeling at peptide level, after protein digestion like iTRAQ and labeling during protein digestion, such as [18]O labeling (Capelo et al. 2010). SILAC can only be used for samples which are produced using labeled amino acids, while the other methods can be used for all types of protein samples. Thiede *et al.* have recently introduced a promising new labeling method with relative or absolute quantification for identification and quantification of two differentially labelled states using MS/MS spectra, and which is called isobaric peptide termini labeling (IPTL) (Thiede and Koehler 2010). The method involve digestion of the protein samples and cross-wise labeling of N- and C-terminal ends of the obtained peptides, like the principle in [18]O labeling (Thiede and Koehler 2010).

The digestion efficiency in comprehensive proteomics is as important as the digestion repeatability in targeted proteomics. Everyone working in this field should strive to have control over these parameters during the sample preparation in proteomics, producing correctly identified proteins and reliable results. The focus in this review is on the in-solution based protocols in comprehensive proteomics, with emphasis on in-solution tryptic digestion and alternative methods to speed up the digestion, and also on how to evaluate the digestion efficiency of the used method.

## 2. Factors influencing proteolytic results

An issue that is little discussed in the literature of proteomics is the sample handling prior to the protein digestion. Some mention the need for enrichment, or elimination of interfering substances (López-Ferrer et al. 2006), but few focus on the steps prior to the enzymatic digestion of the protein fractions.

### 2.1 Protein concentration

A proper digestion procedure starts with the measurement of the protein content of the sample. This is necessary to determine for, among others, the needed amount of reduction and alkylation reagents, as well as the amount of enzyme in in-solution digestion. Quantifying the protein content of a sample separated on a gel is often relatively easy. In this case, guidelines of intensity of the stained gel-band can be used as a "semi-quantitative" measurement. The amount of the total protein content of gel-free samples can be measured with standard procedures like the NanoDrop (detection down to 10-15 µg/ml, using 2 µl sample) (NanoDrop 2011), the Bradford assay (detection down to 2.5 µg/ml, using 150 µl sample) (Bradford 1976) and the BCA assay (detection down to 20 µg/ml, using 25 µl sample) (Smith et al. 1985).

### 2.2 Keratin contamination

Avoiding keratin contamination, which is a problem common in both 1D or 2D gel and in-solution methods, but mostly in the gel-based analysis (Bell et al. 2009), is important. Keratins are naturally occurring structural proteins and appear more often in the sample as interference from the environment rather than from natural abundance. Fingerprints, hair, dead skin flakes, wool clothing, dust and latex gloves are common sources of contaminating keratins (Greenebaum 2011). If keratins are present at concentration levels greater than that of the protein of interest, their abundance will overwhelm the analytical capacity of the LC-MS system and obscure the protein of interest. This is particularly problematic when performing data dependent mass spectrometry, as the peptides from the more abundant keratins will be selected for tandem-MS analysis, providing little or no information about the actual proteins of interest. However, at low concentration levels, compared to the protein of interest, keratins are not a problem at all (Greenebaum 2011).

### 2.3 Detergents

Detergents are often used for total solubilisation of cells and tissues in biochemical studies, and sodium dodecyl sulphate (SDS) is often the choice. However, even at low concentrations, detergents can give rise to problems both concerning enzymatic digestion and in the subsequent LC-MS analysis. Hence it is most often necessary to deplete the detergents prior to the steps in the analytical method hampered by the detergent, or to find alternative ways to lyse the cells which are more compatible with the downstream steps in the analysis. This problem will be further discussed in section 5.2.2.

## 3. From proteins to peptides

### 3.1 Denaturation, reduction and alkylation

Prior to in-solution protein digestion the proteins in most samples need to be denatured, reduced and alkylated, using various reagents, for the proteolytic enzyme to be able to efficiently cleave the peptide chains of the proteins. A sample preparation workflow is presented in Figure 2 together with different suggested procedures to accelerate the tryptic digestion of proteins to peptides. These methods are presented in section 5.

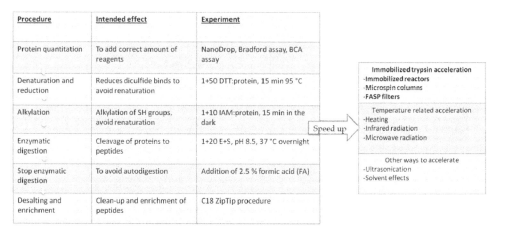

| Procedure | Intended effect | Experiment |
|---|---|---|
| Protein quantitation | To add correct amount of reagents | NanoDrop, Bradford assay, BCA assay |
| Denaturation and reduction | Reduces diculfide binds to avoid renaturation | 1+50 DTT:protein, 15 min 95 °C |
| Alkylation | Alkylation of SH groups, avoid renaturation | 1+10 IAM:protein, 15 min in the dark |
| Enzymatic digestion | Cleavage of proteins to peptides | 1+20 E+S, pH 8.5, 37 °C overnight |
| Stop enzymatic digestion | To avoid autodigestion | Addition of 2.5 % formic acid (FA) |
| Desalting and enrichment | Clean-up and enrichment of peptides | C18 ZipTip procedure |

Speed up

Immobilized trypsin acceleration
-Immobilized reactors
-Microspin columns
-FASP filters

Temperature related acceleration
-Heating
-Infrared radiation
-Microwave radiation

Other ways to accelerate
-Ultrasonication
-Solvent effects

Fig. 2. Procedure, intended effect and experimental conditions of a classical workflow for in-solution based sample preparation approaches in proteomics. To the very right different accelerating digestion techniques are presented.

In a study by Proc *et al.* the denaturation process of human plasma proteins was examined applying 14 different combinations of heat, solvents, chaotropic agents and surfactants for their effectiveness to improve tryptic digestion (Proc et al. 2010). The experiment was performed by quantifying the production of proteolytic tryptic peptides from 45 moderate-to-high-abundance plasma proteins which were grouped into rapidly digested proteins, moderately digested proteins and proteins resistant to digestion. Proc *et al.* did not find an "optimal" digestion method for all 45 proteins, but the denaturation procedure with the surfactant sodium deoxycholate (DOC), which is more compatible with MS than SDS, together with a digestion time of 9 hours, was found to be the most promising protocol for all proteins (Proc et al. 2010).

Denaturation and reduction can often be carried out simultaneously by a combination of heat and a reagent, like 1,4-dithiothreitol (DTT) (Choudhary et al. 2002), β-mercaptoethanol (Sundqvist et al. 2007) or tris(2-carboxyethyl)phosphine (Hale et al. 2004). Most used is DTT, which is a strong reducing agent, that reduce the disulfide bonds and prevent inter and intra-molecular disulfide formation between cysteines in the protein. By combining denaturation and reduction, renaturation of the proteins due to reduction of the disulfide bonds can be avoided (see Figure 3). Renaturation can be a problem using heat solely as the denaturation agent (Strader et al. 2005; Capelo et al. 2009).

Following protein denaturation and reduction, alkylation of cysteine is necessary to further reduce the potential renaturation (Figure 3), and the most commonly used agents for alkylation of protein samples prior to digestion are iodoacetamide (IAM) and iodoacetic acid (IAA) (López-Ferrer et al. 2006; Vukovic et al. 2008).

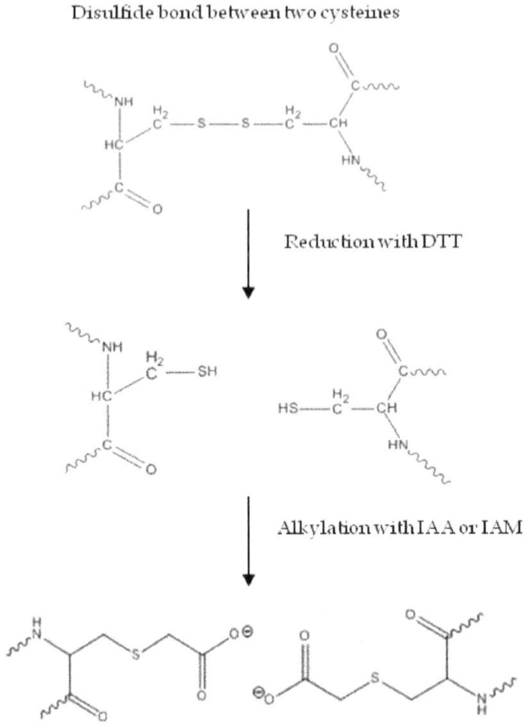

Disulfide bond between two cysteines

Reduction with DTT

Alkylation with IAA or IAM

Fig. 3. The reduction and alkylation process: The breaking of disulfide bonds in proteins. Reduction by DTT to form cysteine residues must be followed by further modification of the reactive –SH groups (to prevent reformation of the disulfide bond) by acetylation by, in this case iodoacetic acid (adapted from (Nelson and Cox 2008)).

## 3.2 Trypsin digestion

Protein digestion with proteases is one of the key sample-preparation steps in proteomics, followed by LC-MS. As already mentioned, trypsin is the most commonly used protease for this purpose since it has a well defined specificity; it hydrolyzes only the peptide bonds in which the carbonyl group is followed either by an arginine (Arg) or lysine (Lys) residue, with the exception when Lys and Arg are N-linked to Aspartic acid (Asp). The cleavage will not occur if proline is positioned on the carboxyl side of Lys and Arg. Since trypsin is a protein it may digest itself in a process called autolysis. However, $Ca^{2+}$, naturally present in most samples, binds at the $Ca^{2+}$-binding loop in trypsin and prevents autolysis (Nord et al. 1956). With the modified trypsin presently used in most laboratories, autolysis is additionally reduced. Still addition of 1 mM $CaCl_2$ is recommended in the digestion medium, but not always absolutely necessary, when the contribution of $Ca^{2+}$ from natural sources is low (Minnesota 2011). Tryptic digestion is performed at an optimal pH in the range 7.5-8.5 (Worthington 2011), and commonly at 37 °C for in-solution digestion. Thus prior to the addition of trypsin, a buffer is added (usually 50 mM triethyl ammonium bicarbonate (tABC) or 12.5 mM ammonium bicarbonate (ABC) buffer (López-Ferrer et al. 2006) to provide an optimal pH for the enzymatic cleavage. A 2-amino-2-hydroxymethyl-propane-1,3-diol (Tris) buffer may also be used for this purpose, but it should be taken into consideration that the Tris buffer is incompatible with the down stream MS analysis, such as MALDI and ESI-MS, and needs to be depleted through solid phase extraction (SPE) or ZipTips prior to such (Shieh et al. 2005; Sigma-Aldrich 2011).

Information about the enzyme to protein ratio needed for digestion of a protein sample is crucial to ensure an enzyme amount sufficient to perform the digestion, but not too high resulting in autolysis products from the trypsin used. Recent experiments indicate that a sufficient ratio of enzyme to substrate (E+S) is 1+20 (Hustoft et al. 2011). For targeted proteomics it may be beneficial to perform a pilot study on the necessary digestion time for the type of sample to be analyzed, to obtain an optimal digestion efficiency of the sample. For more comprehensive proteomics a longer digestion time, up to 9 hours is recommended to ensure the best overall digestion efficiency, as described by Proc *et al.* (Proc et al. 2010). Thus dealing with these long digestion times, an overnight digestion is often more convenient, starting with the post digestion sample preparation steps the following day.

Proteins may act differently in different environments and less effective digestions have been observed when model proteins were digested in a mixture as compared to being digested separately (Hustoft et al. 2011). One reason for these observations could be increased competition for the trypsin cleavage sites, when more proteins are digested together.

As mentioned in the introduction one of the main issues regarding digestion is how to measure the digestion efficiency of a method for a given complex sample of proteins. Examples from the literature show that different groups use various measures for the efficiency of their digestive method, where amino acid SQ % is the most common. However, based on our experience, we question whether the SQ % can serve as a reliable measurement of digestion efficiency, or not? Using a relatively high concentration of 250 ng/ ml of each of the model proteins, no significant difference in SQ % could be seen with a 5 min digestion versus an overnight digestion. Thus another measure for the digestion efficiency had to be evaluated. Since cyt-C was one of the model proteins, undigested intact protein could be detected by the Ion-Trap MS being used. The area of the intact protein peak decreased with increasing digestion time, - indicating better trypsination efficiency. The size of the intact protein peak could hence in some cases be used to compare the efficiency of digestion methods (Hustoft et al. 2011). When exploring the potential of microwave oven accelerated digestion of a mixture of proteins, different temperatures were examined both for the microwave oven and the Thermoshaker control samples. A decrease in the intact protein peak of cyt-C was detected indicating better efficiency at higher temperatures, in both cases. However, the peak area of four distinct tryptic peptides from cyt-C revealed that the decrease in cyt-C peak area was caused by denaturation of the sample as a function of higher temperatures, and not because of increased digestion. Hence, the area of the undigested protein peak is not necessarily a good measure for the digestion efficiency. Another way to describe the digestion efficiency is through the yield of peptides, used to study the effect of temperature, enzyme concentration, digestion time and surface area of the gel pieces in in-gel proteomics (Havliš et al. 2003).

## 4. Sample handling post digestion

### 4.1 Clean-up and enrichment of digests

Prior to LC-MS analysis, the digests must be purified to remove e.g. buffers and salts added during the sample preparation. This is most often carried out with ZipTips, which concentrate and purifies the samples for sensitive downstream analysis (Capelo et al. 2009). A C18 ZipTip is a 10 µl pipette tip with a 0.6 or 0.2 µl bed of C18 silica based medium fixed at its tip, used for single-step desalting, enrichment, and purification. Such ZipTips can be used for purification of, for instance peptides, proteins and nucleic acids. Purifying tryptic peptides with the C18 ZipTip results in high recovery, but noteworthy is that the capacity of the C18 ZipTips is limited; however, up to 10 µg digested protein could be loaded without losses (Hustoft et al. 2011). Another possible disadvantage of the C18 ZipTip procedure is the loss of small hydrophilic peptides which may be lost due to washing with an aqueous mobile phase containing 0.1 % trifluoroacetic acid (TFA) (Hustoft et al. 2011). Still, when the sample must be purified prior to LC-MS analysis the ZipTips are convenient to use because they are easy to handle and commercially available at a reasonable price, producing good recoveries.

## 5. Accelerating the protein digestion

An efficient proteolytic digestion, which is important to correctly identify proteins in comprehensive proteomics and to obtain low detection limits in targeted proteomics, requires the generation of peptides in a minimal amount of time. Conventional methods often involve up to 12-16 hours of incubation, but digestion times up to 24 hours are reported, due to protein heterogeneity in samples (López-Ferrer et al. 2006). Alternative methods have therefore been introduced in order to speed up the digestion method. Capelo *et al.* report eight ways to speed up the protein identification workflow (Capelo et al. 2009); heating, microspin columns, ultrasonic energy, high pressure, infrared (IR) energy, microwave energy, alternating electric fields and microreactors where the trypsin is immobilized on a solid support. The pros and cons of these methods were assembled in a table, including citations or validations of the methods from other research groups (Capelo et al. 2009). Capleo *et al.* found that heating, ultrasonication, microwave energy and microreactors (immobilized trypsin) are used in most applications, and recommend that the systems with microspin columns, high pressures, alternating electric fields and IR energy need to be further validated. In a recent study (Hustoft et al. 2011) we have evaluated some of these techniques; IR energy, microwave energy, solvent effects as well as a newly developed filter aided sample preparation (FASP) technique to perform both depletion of detergents like SDS and tryptic digestion of proteins on the same filter device. The different methods are grouped into "temperature related accelerated digestion", "immobilized trypsin accelerated digestion" and "other ways to accelerate digestion" in the following. The terminology used gives an indication of the acceleration method for enzymatic digestion.

## 5.1 Temperature related accelerated digestion
### 5.1.1 Heating
Enzymes perform best at a given temperature and for in-solution tryptic digestion, 37 °C has been suggested as the optimal temperature (Havliš et al. 2003), and is the temperature most commonly used both for overnight in-gel and in-solution based tryptic digestion. Havlis *et al.* showed that reductive methylation of trypsin decreases autolysis and shifts the optimum of its catalytic activity to 50-60 °C, with enzymatic digestion of bovine serum albumin (BSA) 12 times faster than in-solution at 37 °C, using the yield of peptides as a parameter of the digestion efficiency (Havliš et al. 2003). From time to time some approaches have been introduced regarding the use of elevated temperatures for trypsin digestion (Capelo et al. 2009), but no new papers have been published recently.

### 5.1.2 Ultrasonic assisted digestion

Among the different ways to speed up protein digestion, ultrasonic energy has been considered the most promising method of the techniques requiring specialized equipment (Capelo et al. 2009). Three different commercial devices are used for ultrasonication in laboratories today. The most available is the ultrasonic bath, but for the purpose of accelerating tryptic digestion this is not sufficiently powerful to shorten the digestion times (Capelo et al. 2009). Regardless of this, Li *et al.* claimed that an ultrasound bath-assisted method gave successful in-solution proteolysis of three model proteins; BSA, cyt-C and myoglobin, revealing higher SQ % than conventional overnight incubation at 37 °C (Li et al. 2010) . However, the experimental set up and type of samples used should be considered carefully. It would probably be more correct to compare the ultrasonic bath method to 37 °C incubation without the ultrasonic bath, using proteins denaturated, reduced and alkylated in the same fashion. Sonoreactors and ultrasonic probes are more effective, revealing a higher number of peptides and thus better SQ %, giving digestion in seconds as shown in the direct ultrasonic assisted enzymatic digestion of the soybean proteins (Domínguez-Vega et al. 2010). Carreira *et al.* proposed in another study a methodology that uses ultrasonic energy to speed up the protein digestion and throughput of [18]O labeling for protein quantification and peptide mass mapping through mass spectrometry based techniques (Carreira et al. 2010). This is a promising technique to accelerate the trypsin digestion of proteins, thus requiring specialized equipment, as mentioned in the start of this section.

### 5.1.3 Infrared radiation assisted digestion

In 2008 Wang *et al.* introduced a system where infrared (IR) energy was used to speed up the rate of trypsin digestion of proteins (Wang et al. 2008). The type of instrumentation used is presented in Figure 4, and the infrared light contributed, according to the authors, to shorter digestion times by increasing the excitation of the molecules and thus increasing the interaction between trypsin and the peptide bonds in the molecule.

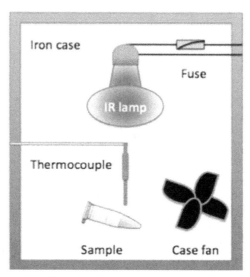

Fig. 4. Schematic diagram of the IR-assisted proteolysis system, adapted from (Wang et al. 2008).

In their first study IR assisted digestion was carried out for 5 min with trypsin in-solution and revealed almost a 100 % increase in SQ % of BSA and a 20 % increase in the SQ % of myoglobin compared to conventional trypsin digestion for 12 h at 37 °C. The method was considered repeatable when examined with a series of eight digestions giving myoglobin SQs of 90 % for all. Wang *et al.* later used the same system to study the digestion by another commonly used protease, α -chymotrypsin, which typically needs in-solution digestion times of 12-24 h (Wang et al. 2008). Using IR radiation the digestion time was reduced to 5 min for the digestion of BSA and cyt-C with SQs of 41 and 75 %, respectively. When the IR contribution was eliminated, the SQs were reduced to 11 and 56 % for BSA and cyt-C, respectively. For comparison the 12 h digestion at 37 °C yielded SQs comparable to those of 5 min IR radiation (37 % (BSA) and 75 % (cyt-C)). The same system was further examined three times in the years 2008-2009 (Bao et al. 2008; Wang et al. 2008; Bao et al. 2009) for the digestion of proteins on-plate MALDI-TOF-MS for in-gel proteolysis and one approach using trypsin-immobilized silica microspheres for peptide mapping. In 2010 another technique called photo thermal heating was introduced by Chen *et al.* A near infrared (NIR) diode laser was used to increase the reaction temperature during tryptic digestion on a Glass@AuNP slide, in a short period of time. The technique was used for four different proteins without the need for reduction and alkylation. The sequence coverages were in the range 43-95 % compared to 28-75 % with 12 h incubation at 37°C (Chen et al. 2010). Unfortunately no comparison of trypsin digestion efficiency with and without the NIR source was undertaken. We have found that, proteins can be digested in an IR oven, but compared to the traditional digestion procedure using 37 °C, there are no indications that the IR method has improved digestion efficiency for the commonly employed amount of proteins, at digestion times from 5 minutes up to 5 hours (Hustoft et al. 2011).

### 5.1.4 Microwave assisted digestion

Microwave assisted tryptic digestion was introduced in 2002 by Pramanik *et al.* as a tool to speed up the proteolytic cleavage of proteins (Pramanik et al. 2002). Other enzymes, as the endoproteinase Glu-C, has been reported to be inactivated by microwave induced denaturation, but trypsin digestion is accelerated according to the authors (Lill et al. 2007). In an attempt to investigate the acceleration of enzymatic cleavage, trypsin digestion with unmodified trypsin was performed at different microwave temperature settings, 37, 45 and 55 °C. The temperatures in the sample were found to be significantly higher than their microwave settings, and the authors emphasized that it was important to note the elevation of the reaction temperature which greatly enhanced the digestion reaction (Pramanik et al. 2002). Whether microwave accelerated digestion is a convenient way of heating, or whether the microwaves have a non-thermal positive effect on the digestion reaction, can be questioned. In a review on microwave-assisted proteomics, Lill *et al.* addressed the "heating principle" and stated that the kinetics in the microwave assisted incubation are different from the water bath incubation in that proteolysis was greatly enhanced when mediated by microwave radiation and that tightly folded proteins benefit the most from the microwave-assisted proteolysis (Lill et al. 2007). Two papers by Lin *et al.* (Lin et al. 2007; Lin et al. 2008) and one by Hahn *et al.* (Hahn et al. 2009) showed acceleration of digestion through a combination of immobilized trypsin and microwave radiation, when the digestion efficiency was measured as SQ % of different model proteins. In a short communication by Reddy *et al.* various solvents, temperatures and different enzyme: substrate (E+S) ratios were compared to see how they affected protein digestion under conventional heating and microwave-assisted digestion. Digestion efficiencies were referred to as the ratio of the abundance of the most abundant peptide product to that of this peptide plus the undigested protein. Optimal conditions were found to be microwave-assisted irradiation at 60 °C for 30 min in a 50 mM Tris buffer with a of 1:5 or 1:25 (Reddy et al. 2010). It should be noted that this method is incompatible with subsequent MS analysis when Tris is used as a buffer, without a buffer exchange. To make sure that no denaturation of the trypsin occurs at 60 °C, modified enzyme should be used.

The microwave approach has also been used in some recent papers (Hasan et al. 2010; Liu et al. 2010), for effective enrichment of phosphopeptides and $^{18}O$ labeling. High sensitivity and SQ % of phosphopeptides were obtained and explained by absorption of microwave radiation by accelerated activation of trypsin for efficient digestion of the phosphoproteins (Hasan et al. 2010). The microwave assisted $^{18}O$ labeling resulted in peptide mixtures with $^{18}O$ incorporation in less than 15 min with a low rate of back exchange (Liu et al. 2010). We have evaluated microwave assisted protein digestion using both a specialized temperature controlled microwave oven and a domestic microwave oven. No differences in SQ % (or area of intact protein peak of cyt-C) were found for microwave and temperature assisted protein digestion for four model proteins. As previously suggested, microwave irradiation seems to have no advantage over normal temperature induced digestion, within our experimental framework (Hustoft et al. 2011).

## 5.2 Immobilized trypsin accelerated digestion
### 5.2.1 Microreactors

The immobilization of enzymes onto solid materials can be traced back to the 1950s according to Ma *et. al* (Ma et al. 2009), and in the last decades numerous immobilization methods have been developed. Proteolytic enzymes can be covalently bonded or physically adsorbed onto different carriers, such as inorganic silica materials, and organic materials that display a great variability and good biocompatibility like polystyrene divinylbenzenes (PS-DVB), polyacrylamides and methacrylates (Ma et al. 2009). These types of reactors appear to have a promising future, and constitute the most used accelerating digestion techniques the last couple of years. Immobilized microreactors have a high enzymatic turnover rate, low reagent consumption, less contribution of enzyme autolysis and the possibility to be coupled on-line to nanoLC-MALDI or nanoLC-ESI (Capelo et al. 2009; Ma et al. 2009). In a review by Monzo *et al.* from 2009 the most important proteolytic enzyme-immobilization processes are summarized with emphasis on trypsin immobilized micro- and nanoreactors (Monzo et al. 2009). Another review on immobilized enzymatic reactors was published by Ma *et al.* in 2009 (Ma et al. 2009). Different inorganic and organic carriers for particle based, monolithic, open tubular capillaries and membranes with immobilized enzymes were included and the authors predicted that immobilized enzyme reactors might be one of the key points to combine the top-down and bottom-up strategies in the field of proteomics. Still, some characteristics like higher mechanical strength, larger surface area, lower backpressure, higher enzyme loading capacity and better biocompatibility, are needed.

In 2009 three papers concerning immobilized enzyme microreactors and LC-MS/MS were published (Krenkova et al. 2009; Yamaguchi et al. 2009; Yuan et al. 2009). Yuan *et al.* presented an integrated protein analysis platform based on column switch recycling size exclusion chromatography (SEC), a microenzymatic reactor and μLC-ESI-MS/MS. The system combines conventional SEC separation of intact proteins with on-line protein digestion on an immobilized enzymatic reactor (IMER) of conventional size and subsequent separation of peptides on a 300 μm (inner diameter) ID C18 column using ESI-MS/MS for identification (Yuan et al. 2009). The system requires large sample amounts and needs to be evaluated with real samples in order to be classified as a promising tool in proteomic studies. Monolithic enzymatic microreactors have been applied to digest, among others, immunoglobulin G at room temperature in only 6 minutes with reduced nonspecific adsorption of proteins and peptides to the stationary phase, as shown by Krenkova et al. (Krenkova et al. 2009). The SQ % was used as a measure for the digestion efficiency. Another microreactor was introduced by Yamaguchi et al., using a PTFE microtube (500 μm ID, 13 cm length) with covalent binding of the enzyme. This tube was used to digest cyt-C and BSA, where the proteins (denatured in guanidine-HCl) were pumped into the immobilized microreactor and the tryptic peptides were subsequently purified on a C18 cartridge prior to LC or MS analysis. Yamaguchi et al. claimed that BSA could be digested without any reduction and alkylation procedures. Immobilized assisted digestion for 5.2 min at 30°C was compared with in-solution digestion of denatured BSA for 15 h at 37°C, however producing a rather low SQ % in both cases, 12 and 8 %, respectively (Yamaguchi et al. 2009). The authors claim that the low SQ % obtained was due to the stabilized tertiary structure by the 16 disulphide bonds that BSA contains, and probably better results would be obtained if reduction and alkylation of BSA had been performed in advance.

Xu et al. demonstrated a microporous reactor where polystyrene sulfonate and trypsin were adsorbed to a nylon membrane, to make a syringe based system for protein digestion. They used SQ % as the parameter for digestion efficiency claiming that the sequence coverage is a function of both the completeness of the protein digestion and the detection efficiency for the various tryptic peptides (Xu et al. 2010). The system showed improved SQ % of 84 % for BSA in only 6.4 seconds residence time compared to in-solution digestion for 16 h, and more promising cleavage in the presence of small amounts of SDS (Xu et al. 2010). Recently a critical overview of some highly efficient immobilized enzyme reactors termed IMERs, were presented (Ma et al.). This paper includes some newly developed IMERs and systems for protein-expression profiling, IMERs for characterization of proteins with PTMs and IMERs for protein quantification.

There are some drawbacks associated with the use of microreactors, like for instance the costs of the commercially available products of immobilized reactors. Self-fabrication requires adequate tools and experience in immobilization on different supports with enzymes. Automation is also still not easily achieved. However, as previously mentioned, on-going research can be expected to improve the techniques.

### 5.2.2 From microspin columns to filter-aided sample preparation

Commercial microspin columns or so called trypsin spin columns, where trypsin is immobilized at a high density on a solid support, - has been introduced by among others Sigma-Aldrich. It has been claimed that they reduce digestion times of proteins to 15 min, compared to conventional digestion times of 12 h, and give little autolysis fragments. However, the total microspin column method has been found to be both labor intensive and complex. The disappearance of these columns have been predicted because they do not give any apparent advantage over other types of immobilized trypsin which are commercially available and can be prepared in any lab (Capelo et al. 2009). This prediction can also be supported by the fact that Sigma-Aldrich's trypsin columns are no longer available. A kit intended for ¹⁸O labeling called Prolytica ¹⁸O labeling kit from Stratagene, based on trypsin immobilized spin columns, is also now out of production. Promega additionally had one product available, called "Immobilized trypsin", where, with the use of the spin column format, digested peptides could easily be separated from the immobilized trypsin, reducing enzyme interference during analysis (Wiśniewski et al. 2010). This product is also now withdrawn because of low demand.

Wisniewski *et al.* presented a "Universal sample preparation method for proteome analysis" based on a Filter-Aided Sample Preparation (FASP) (Wisniewski et al. 2009). The enzyme is not directly immobilized onto the ultrafiltration device, but the device acts as a "proteomic reactor" for detergent removal, buffer exchange, chemical modification and protein digestion, where trypsin is added in a dissolved form to the filter (Figure 5). Lately four other papers have been published based on this method, and it seems promising for both membrane proteins, brain phosphoproteins and the N-glycoproteins (Wisniewski et al. 2009; Ostasiewicz et al. 2010; Wisniewski et al. 2010; Zielinska et al. 2010).

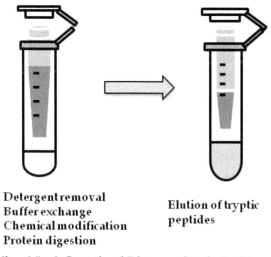

Detergent removal
Buffer exchange
Chemical modification
Protein digestion

Elution of tryptic
peptides

Fig. 5. The Amicon Ultra-0.5 mL Centrifugal Filters used in the FASP procedure. Adapted from, (Millipore 2011).

We, however, found that the filter device was not able to deplete all SDS, and this can lead to problems with the subsequent LC-MS analysis (Hustoft et al. 2011). The FASP procedure was found rather time consuming (using up to 3.5 h prior to the trypsin digestion) and the recommended 1:100 enzyme to protein ratio was not found satisfactory in our laboratory. Recently the FASP method was made commercially available through a FASP™ Protein Digestion Kit, from Protein Discovery. In this protocol the time of the centrifugation steps has been decreased, still it takes more than 2 hours to complete the protocol prior to 4-18 hours of trypsin digestion. Since the method has been made commercially available through a kit, and found to be convenient (Ostasiewicz et al. 2010; Wisniewski et al. 2010; Zielinska et al. 2010; Hustoft et al. 2011) this method of trypsin digestion can be recommended when e.g. working with in-solution digestion of samples solubilised in detergents like SDS.

## 5.3 Solvent effects

The enzyme activity can also be improved in organic solvents as reported by Gupta and Roy (Gupta and Roy 2004). This was additionally shown by Strader *et al.* (Strader et al. 2005) who used an organic-aqueous system for digestion, containing 80 % acetonitrile (ACN), and which consistently provided the most complete digestion of microgram to nanogram quantities of proteins, by producing more peptide identifications at a shorter time (only 1 h compared to overnight). In a following paper Hervey *et al.* compared five different in-solution digestion protocols revealing that by adding 80 % ACN to the digestion solution the sequence coverages were as good as or in some cases better than using solvents with lower ACN % or chaotropes in the digests (Hervey IV et al. 2007). Addition of ACN to the digestion medium can cause (partial) denaturation of proteins and thus better accessibility to the cleavage sites of the protein. ACN can also improve digestion efficiency and enhance the solubility as well as elution of tryptic digests from e.g. a trypsin immobilized column (Tran et al. 2008). For the digestion of cyt-C, BSA, lysozyme and α-lactalbumin, addition of organic solvent up to 80 % did not increase the digestion efficiency regarding the area of the intact cyt-C peak or increased the sequence coverage (Hustoft et al. 2011). When more than 40 % ACN was used in combination with the tABC buffer, protein precipitation was seen. A solution to this problem is to use a buffer system with Tris-HCl/CaCl$_2$ when amounts of 40 % organic solvent or more are added to the sample solution (Hustoft et al. 2011). But, as before mentioned the Tris buffer is incompatible with the subsequent MS analysis and needs to be depleted prior to such.

## 6. Conclusions and recommended trypsin digestion procedure for LC-MS based proteomics

As has been pointed out, for some of the accelerating techniques used for tryptic digestion there is a need for more validation. We have thoroughly evaluated four of these techniques (Hustoft et al. 2011) finding no clear increase in the digestion efficiency (measured as SQ % or intact protein peak of cyt-C) of four model proteins when using IR energy, microwave energy, aqueous-organic solvent systems or FASP filters. What is of importance when comparing novel methods to established ones, is to include control experiments where the same treatment is used but without the accelerating factor for the control. When the digestion efficiency is measured based on amino acid sequence coverage, results have been found to be strongly dependent on the LC-MS data quality of the analyzed samples. Hence more replicates are strongly recommended for correct evaluation of the methods. The MS instrument available is of importance for examining the digestion efficiency and also the choice of model proteins are crucial because of their different response to tryptic digestion. Working with conventional shotgun (bottom-up) proteomic techniques the overall digestion efficiency is more intricate to study than when working with targeted proteomics. In targeted proteomics much more information can be found about the proteins to be determined, e.g. whether they have cysteines and need to be reduced and alkylated prior to enzymatic digestion. The literature can be searched in order to find relevant information and even established methods used for the targeted proteins, and selected reaction monitoring (SRM) can be used for targeted quantitative proteomics.

It should be kept in mind however, that many different variants of key words denoting the same method or process are used in the literature. One example is the method of trypsin digestion where different papers were found depending on which key word was entered

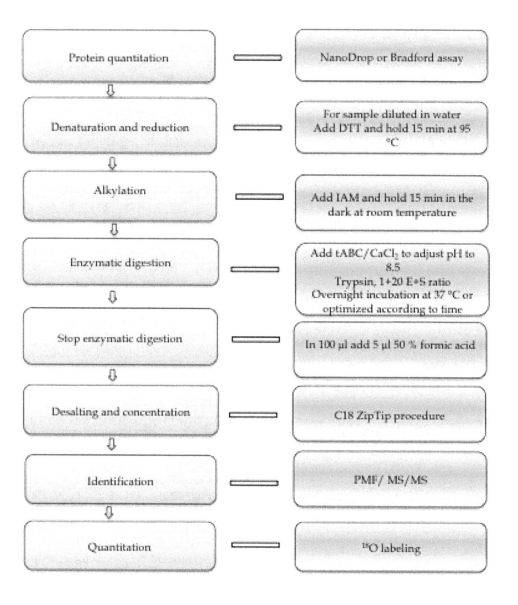

Fig. 6. A recommended procedure for in-solution based sample preparation and protein identification. PMF refers to peptide mass fingerprinting. (PFM).

into the search field of, in this case, SciFinder®: *Proteolysis, protein digestion, trypsin digestion, tryptic digestion, enzyme reaction, enzymatic digestion* or *enzymatic cleavage*, all produced hits for papers concerning trypsin/tryptic digestion (as we have chosen to use). Rounding up with Mitchell, he refers to a test sample study done in 2009, where 27 labs were included in reproducibility testing of standardized samples of 20 known proteins each containing one or more unique tryptic peptide. Only seven of the 27 labs reported the 20 proteins correctly, and only one identified all the proteolytic peptides (Bell et al. 2009). When they collected and analyzed the raw MS data from the labs, they found that all proteins and most peptides had been detected in all labs, but just not been interpreted correctly, indicating that it was the human element that failed. Due to the difficulties in correctly identifying the proteins in comprehensive proteomics the future of the field of proteomics will probably be more directed against targeted proteomics. However, as mentioned in the introduction, not every proteomic problem can be solved through targeted proteomics and it will still be a need for comprehensive analyses of complex samples.

Reviewing and in-house experiments of some of the suggested accelerating methods for trypsin digestion did not provide us with a better procedure for speeding up the sample preparation step in in-solution based proteomics, with the possible exception of ultrasonication. A complete recommended sample preparation procedure for newcomers in the field is presented in Figure 6, partially based on some of the conclusions from our investigations. The recommended procedure gives a robust and effective sample preparation guideline to comprehensive in-solution trypsin digestion of complex protein samples in proteomics. This procedure is more or less business as usual, since none of the suggested accelerating procedures revealed faster and more efficient digestion of proteins, than the inexpensive overnight in-solution digestion at 37 °C.

## 7. References

Anderson, N. and N. Anderson (1998). "Proteome and proteomics: new technologies, new concepts, and new words." *Electrophoresis* 19(11): 1853-1861.

Bao, H., et al. (2008). "Efficient In-Gel Proteolysis Accelerated by Infrared Radiation for Protein Identification." *Journal of Proteome Research* 7(12): 5339-5344.

Bao, H., et al. (2009). "Infrared-assisted proteolysis using trypsin-immobilized silica microspheres for peptide mapping." *PROTEOMICS* 9(4): 1114-1117.

Bell, A. W., et al. (2009). "A HUPO test sample study reveals common problems in mass spectrometry-based proteomics." *Nat Meth* 6(6): 423-430.

Bradford, M. M. (1976). "A rapid and sensitive method for the quantitation of microgram quantities of protein utilizing the principle of protein-dye binding." *Analytical Biochemistry* 72(1-2): 248-254.

Capelo, J., et al. (2010). "Latest developments in sample treatment for 18O-isotopic labeling for proteomics mass spectrometry-based approaches: A critical review." *Talanta* 80(4): 1476-1486.

Capelo, J. L., et al. (2009). "Overview on modern approaches to speed up protein identification workflows relying on enzymatic cleavage and mass spectrometry-based techniques." *Analytica Chimica Acta* 650(2): 151-159.

Carreira, R. J., et al. (2010). "Indirect ultrasonication for protein quantification and peptide mass mapping through mass spectrometry-based techniques." *Talanta* 82(2): 587-593.

Chen, J.-Y., et al. (2010). "Multilayer gold nanoparticle-assisted protein tryptic digestion in solution and in gel under photothermal heating." *Analytical and Bioanalytical Chemistry*: 1-9.

Choudhary, G., et al. (2002). "Multiple Enzymatic Digestion for Enhanced Sequence Coverage of Proteins in Complex Proteomic Mixtures Using Capillary LC with Ion Trap MS/MS." *Journal of Proteome Research* 2(1): 59-67.

Domínguez-Vega, E., et al. (2010). "First approach based on direct ultrasonic assisted enzymatic digestion and capillary-high performance liquid chromatography for the peptide mapping of soybean proteins." *Journal of Chromatography A* 1217(42): 6443-6448.

Greenebaum, C. C. (2011). From http://www.umgcc.org/research/proteomics_services.htm.

Gupta, M. N. and I. Roy (2004). "Enzymes in organic media." *European Journal of Biochemistry* 271(13): 2575-2583.

Hahn, H. W., et al. (2009). "Ultrafast Microwave-Assisted In-Tip Digestion of Proteins." *Journal of Proteome Research* 8(9): 4225-4230.

Hale, J. E., et al. (2004). "A simplified procedure for the reduction and alkylation of cysteine residues in proteins prior to proteolytic digestion and mass spectral analysis." *Analytical Biochemistry* 333(1): 174-181.

Hasan, N., et al. (2010). "Two-step on-particle ionization/enrichment via a washing- and separation-free approach: multifunctional $TiO_2$ nanoparticles as desalting, accelerating, and affinity probes for microwave-assisted tryptic digestion of phosphoproteins in ESI-MS and MALDI-MS: comparison with microscale $TiO_2$" *Analytical and Bioanalytical Chemistry* 396(8): 2909-2919.

Havliš, J., et al. (2003). "Fast-Response Proteomics by Accelerated In-Gel Digestion of Proteins." *Anal. Chem.* 75(6): 1300-1306.

Hervey IV, W., et al. (2007). "Comparison of digestion protocols for microgram quantities of enriched protein samples." *J. Proteome Res* 6(8): 3054-3061.

Hustoft, H. K., et al. (2011). "Critical assessment of accelerating trypsination methods." *J. Pharmaceut. Biomed. Anal.* 10.1016/j.jpba.2011.08.013.

IS, M. S. R. (2011). from http://ionsource.com/.

Krenkova, J., et al. (2009). "Highly Efficient Enzyme Reactors Containing Trypsin and Endoproteinase LysC Immobilized on Porous Polymer Monolith Coupled to MS Suitable for Analysis of Antibodies." *Anal. Chem.* 81(5): 2004-2012.

Li, Y.-P., et al. (2010). "Ultrasound-assisted urea-free chemical denaturants combined with thermal denaturation to accelerate enzymatic proteolysis." *Fenxi Huaxue* 38(Copyright (C) 2010 American Chemical Society (ACS). All Rights Reserved.): 663-667.

Lill, J. R., et al. (2007). "Microwave-assisted proteomics." *Mass Spectrometry Reviews* 26(5): 657-671.

Lin, S., et al. (2007). "Development of microwave-assisted protein digestion based on trypsin-immobilized magnetic microspheres for highly efficient proteolysis followed by matrix-assisted laser desorption/ionization time-of-flight mass spectrometry analysis." *Rapid Communications in Mass Spectrometry* 21(23): 3910-3918.

Lin, S., et al. (2008). "Fast and Efficient Proteolysis by Microwave-Assisted Protein Digestion Using Trypsin-Immobilized Magnetic Silica Microspheres." *Anal. Chem.* 80(10): 3655-3665.

Liu, N., et al. (2010). "Microwave-Assisted 18O-Labeling of Proteins Catalyzed by Formic Acid." *Anal. Chem.* 82(21): 9122-9126.

López-Ferrer, D., et al. (2006). "Sample treatment for protein identification by mass spectrometry-based techniques." *TrAC Trends in Analytical Chemistry* 25(10): 996-1005.

Ma, J., et al. "Immobilized enzyme reactors in proteomics." *TrAC Trends in Analytical Chemistry* In Press, Accepted Manuscript.

Ma, J., et al. (2009). "Recent advances in immobilized enzymatic reactors and their applications in proteome analysis." *Analytica Chimica Acta* 632(1): 1-8.

Millipore. (2011). from http://www.millipore.com.

Minnesota, U. (2011). from http://www.cbs.umn.edu/msp/protocols/insolution.shtml.

Mitchell, P. (2010). "Proteomics retrenches." *Nat Biotech* 28(7): 665-670.

Monzo, A., et al. (2009). "Proteolytic enzyme-immobilization techniques for MS-based protein analysis." *TrAC Trends in Analytical Chemistry* 28(7): 854-864.

NanoDrop (2011).

Nelson, D. N. and M. M. Cox, Eds. (2008). *Lehninger: Principles of Biochemistry*, Freeman and Company.

Nord, F. F., et al. (1956). "On the mechanism of enzyme action. LXI. The self digestion of trypsin, calcium-trypsin and acetyltrypsin." *Archives of Biochemistry and Biophysics* 65(1): 120-131.

Ostasiewicz, P., et al. (2010). "Proteome, Phosphoproteome, and N-Glycoproteome Are Quantitatively Preserved in Formalin-Fixed Paraffin-Embedded Tissue and Analyzable by High-Resolution Mass Spectrometry." *Journal of Proteome Research* 9(7): 3688-3700.

Pramanik, B. N., et al. (2002). "Microwave-enhanced enzyme reaction for protein mapping by mass spectrometry: A new approach to protein digestion in minutes." *Protein Science* 11(11): 2676-2687.

Proc, J. L., et al. (2010). "A Quantitative Study of the Effects of Chaotropic Agents, Surfactants, and Solvents on the Digestion Efficiency of Human Plasma Proteins by Trypsin." *Journal of Proteome Research* 9(10): 5422-5437.

Reddy, P. M., et al. (2010). "Digestion Completeness of Microwave-Assisted and Conventional Trypsin-Catalyzed Reactions." *Journal of the American Society for Mass Spectrometry* 21(3): 421-424.

Shieh, I. F., et al. (2005). "Eliminating the Interferences from TRIS Buffer and SDS in Protein Analysis by Fused-Droplet Electrospray Ionization Mass Spectrometry." *Journal of Proteome Research* 4(2): 606-612.

Sigma-Aldrich. (2011). from
    http://www.sigmaaldrich.com/etc/medialib/docs/Sigma/Bulletin/t6567bul.Par.
    0001.File.tmp/t6567bul.pdf.
Smith, P. K., et al. (1985). "Measurement of protein using bicinchoninic acid." *Analytical Biochemistry* 150(1): 76-85.
Steen, H. and M. Mann (2004). "The abc's (and xyz's) of peptide sequencing." *Nat Rev Mol Cell Biol* 5(9): 699-711.
Strader, M. B., et al. (2005). "Efficient and Specific Trypsin Digestion of Microgram to Nanogram Quantities of Proteins in Organic–Aqueous Solvent Systems." *Anal. Chem.* 78(1): 125-134.
Sundqvist, G., et al. (2007). "A general, robust method for the quality control of intact proteins using LC-ESI-MS." *Journal of Chromatography B* 852(1-2): 188-194.
Thiede, B. and C. Koehler (2010). Mass spectrometry-based quantitative proteomics using isobaric peptide termini labeling, Universitetet i Oslo: 47pp.
Tran, B. Q., et al. (2008). "On-Line multitasking analytical proteomics: How to separate, reduce, alkylate and digest whole proteins in an on-Line multidimensional chromatography system coupled to MS." *J. Sep. Sci.* 31(16-17): 2913-2923.
Vaudel, M., et al. (2010). "Peptide and protein quantification: A map of the minefield." *PROTEOMICS* 10(4): 650-670.
Vukovic, J., et al. (2008). "Improving off-line accelerated tryptic digestion:: Towards fast-lane proteolysis of complex biological samples." *Journal of Chromatography A* 1195(1-2): 34-43.
Wang, S., et al. (2008). "Infrared-Assisted On-Plate Proteolysis for MALDI-TOF-MS Peptide Mapping." *Anal. Chem.* 80(14): 5640-5647.
Wang, S., et al. (2008). "Efficient Chymotryptic Proteolysis Enhanced by Infrared Radiation for Peptide Mapping." *Journal of Proteome Research* 7(11): 5049-5054.
Wang, S., et al. (2008). "Infrared-assisted tryptic proteolysis for peptide mapping." *PROTEOMICS* 8(13): 2579-2582.
Wisniewski, J., et al. (2009). "Universal sample preparation method for proteome analysis." *Nat Meth* 6(5): 359-362.
Wiśniewski, J. R., et al. (2010). "Brain Phosphoproteome Obtained by a FASP-Based Method Reveals Plasma Membrane Protein Topology." *Journal of Proteome Research* 9(6): 3280-3289.
Wiśniewski, J. R., et al. (2009). "Combination of FASP and StageTip-Based Fractionation Allows In-Depth Analysis of the Hippocampal Membrane Proteome." *Journal of Proteome Research* 8(12): 5674-5678.
Worthington, B. C. (2011). from http://www.worthington-biochem.com/try/default.html.
Xu, F., et al. (2010). "Facile Trypsin Immobilization in Polymeric Membranes for Rapid, Efficient Protein Digestion." *Anal. Chem.*: null-null.
Yamaguchi, H., et al. (2009). "Rapid and efficient proteolysis for proteomic analysis by protease-immobilized microreactor." *Electrophoresis* 30(18): 3257-3264.
Yuan, H., et al. (2009). "Integrated protein analysis platform based on column switch recycling size exclusion chromatography, microenzymatic reactor and [mu]RPLC-ESI-MS/MS." *Journal of Chromatography A* 1216(44): 7478-7482.

Zielinska, D. F., et al. (2010). "Precision Mapping of an In Vivo N-Glycoproteome Reveals Rigid Topological and Sequence Constraints." *Cell* 141(5): 897-907.

# 4

# Proteomic Analyses of Cells Isolated by Laser Microdissection

Valentina Fiorilli[1], Vincent P. Klink[2] and Raffaella Balestrini[1]*

*[1]Istituto per la Protezione delle Piante del CNR*
*and Dipartimento Biologia Vegetale*
*dell'Università di Torino, Torino*
*[2]Department of Biological Sciences, Harned Hall,*
*Mississippi State University - Mississippi State*
*[1]Italy*
*[2]USA*

## 1. Introduction

Living organisms conduct biological processes by transducing biotic and abiotic stimuli through gene regulation into well-orchestrated growth and development. Analyzing RNA from a transcribed genome (the transcriptome) is fairly easy due to the availability of various nucleotide sequencing technologies. The translation of a transcriptome provides a blueprint of tens of thousands of different proteins that is known as the proteome (Wasinger et al., 1995). The analysis of proteins from whole tissues, organs or organisms has been made easier thanks to various technologies, including the Edman degradation technique, which is used in sequencing polypeptides (Edman, 1950), and 2 dimensional gel electrophoresis (2-DE), which could resolve up to 10,000 polypeptides (Barrett & Gould, 1973; O'Farrell, 1975; O'Farrell et al., 1977). Thus, the problems of assaying a transcriptome and a proteome from samples isolated from tissues, organs or whole organisms have largely been overcome. However, although it is fairly easy to assay a transcriptome and a proteome at these higher levels of biological organization, assaying a proteome at a cellular resolution level involves a set of problems that in primarily centered on the ability to collect sufficient cells for meaningful studies. The central focus of this chapter is a discussion on the technologies that have allowed the proteomic analyses of cells, isolated from complex samples thanks to a procedure that was first called laser microdissection (Isenberg et al., 1976).

## 2. The diversity of genome activity

Thanks to the recent advances in high-throughput technologies, the past decade has witnessed an explosion of global transcriptome profiling studies, which have produced novel insights into many developmental, physiological and medicinal aspects. Although a great deal of information can be obtained from transcriptome profiling, it is however insufficient for a comprehensive delineation of biological systems. A single approach cannot fully

---

* Corresponding Author

unravel the complexity of living organisms (Persidis, 1998). In addition, enzymatic reactions and signaling pathways depend on the activity of proteins, and protein quantities are regulated by protein synthesis and degradation. These processes may be independent of transcriptional control or have only a weak correlation (Lu et al., 2007; Nie et al., 2007). By generating information on the proteome at cellular resolution, a greater understanding of biological complexity is gained, including post-translational modification, isoforms, and splice variants, which may lead to the identification of important cell-specific protein entities (Schulze & Usadel, 2010). The proteomics approach can shed light on a number of protein species that can be translated from a single gene as a result of alternative splicing (AS) or PTMs. Proteomics analyses can also provide the biological meaning of each variant (Kim et al., 2007; Witze et al., 2007). For example, the *Drosophila Dscam1* gene, which encodes a membrane receptor protein, has 115 exons. The various combinations permit the possibility of 38,016 different proteins to be produced and many have been identified (Schmucker et al., 2000; Chen et al., 2006; Meijers et al., 2007). On the basis of large-scale EST-cDNA alignments and bioinformatics analyses on the genomes of *Arabidopsis thaliana* (thale cress) and *Oryza sativa* (rice), it has been estimated that approximately 30–35% of the their genes are alternatively spliced (Cambpell et al., 2006; Xiao et al., 2005), while in humans up to 95% of multi-exon genes undergo alternative splicing (Pan et al., 2008). The number of alternatively spliced genes in plants is still likely to be underestimated because of the relatively low EST coverage and depth of sequencing of many plant transcripts (Simpson et al., 2008; Xiao et al., 2005). Extensive AS variation has been shown in some *Arabidopsis*-specific gene families, for example in genes encoding serine/arginine-rich proteins, and this results in a five-fold increase in transcriptome complexity (Palusa et al., 2007; Tanabe et al., 2007). In addition, stress conditions seem to dramatically alter the splicing pattern of many plant genes (Ali & Reddy, 2008). For these reasons, there is growing interest in complementing transcriptomic studies with proteomics, which should be considered as part of a multidisciplinary integrative analysis that extend from the gene to the phenotype through proteins.

## 3. Developmental plasticity of protein complexes

Many processes and structures are composed of protein complex aggregates. Protein complexes can vary in size and composition, and range from mega-Dalton assemblies of dozens of proteins (such as the ribosome and the spliceosome) to smaller clusters of just a few proteins. The composition and stability of protein complexes is highly regulated in both a context dependent manner, such as cell-type-specific differences, and a time-dependent manner (Michnick et al., 2004). This biological variability of proteins and their range of physicochemical properties reflect the difficulty of characterizing the structure and the function of protein complexes (Cravatt et al., 2007). In addition, in proteomics, the sample amount is often a limiting factor since, unlike transcript profiling, proteomic approaches cannot benefit from amplification protocols. It should be evident that sensitivity, resolution and speed in data capture are all significant problems with proteomics techniques. In order to circumvent these problems, methods have been developed to extract, separate, detect and identify a wide range of proteins from small sample amounts (Gutstein & Morris, 2007). Technical advances in mass spectometry have facilitated major progress in both the qualitative and quantitative analysis of proteins (Kaspar et al., 2010). Most of these improvements have occurred over the last decade and proteomics has developed a broad range of new protocols, platforms and workflows.

## 4. Proteomic workflow

The workflow of a standard proteomics experiment is crucial for the success of any experiment and it usually includes a good experimental design, an appropriate extraction/fractionation/purification protocol that considers the needs of different samples (tissue/cells or organelle), a suitable separation protocol, protein identification, statistical analysis and validation. The use of proteomics in plant biology research has increased significantly over the last few years with an improvement in both quality and quantitative analysis, inaugurating a new phase of "Second Generation Plant Proteomics" (Jorrín et al., 2009). This growing interest in plant proteomics has continually produced a large number of developmental studies on plant cell division, elongation, differentiation, and formation of various organs using various proteomics approaches (Hochholdinger et al., 2006; Takàč et al., 2011, Miernyk et al., 2011). Most of the studies published in the plant field concern the proteome of *Arabidopsis* and rice. The work has focused on profiling organs, tissues, cells, and/or subcellular proteomes (Rossignol et al., 2006; Komatsu et al., 2007; Jorrin et al., 2007; Jamet et al., 2008; Baerenfaller et al., 2008; Jorrin et al., 2009; Agrawal & Rakwal, 2011) and studying developmental processes and responses to biotic (Mehta et al., 2008) and abiotic stresses (Nesatyy & Suter 2008) using differential expression strategies. However, proteomics research results have recently appeared on several non-model herbaceous non-crop species, woody plants, fruit and forest trees (Table 1). Furthermore, over the past year, proteome analysis has increasingly been applied to the study of cereal grains with the aim of

| model species | | |
|---|---|---|
| *Lycosersicon esculentum* | tomato | Sheoran 2007 |
| *Hordeum vulgare* | wheat | Song et al., 2007 |
| *Glycine max* | soybean | Djordjevic et al., 2007 |
| *Zea mays* | maize | Dembinsky et al., 2007 |
| *Medicago truncatula* | alfalfa | de Jong et al., 2007 |
| **non-model species** | | |
| *Elymus elongatum* | wheatgrass | Gazanchian et al., 2007 |
| *Nicotiana alata* | jasmine tobacco | Brownfield et al., 2007 |
| *Boea hygrometrica* | | Jiang et al., 2007 |
| *Xerophyta viscose* | | Ingle et al., 2007 |
| *Solanum chacoense* | chaco potato | Vyetrogon et al., 2007 |
| *Citrullus lanatus* | wild watermelon | Yoshimura et al., 2008 |
| *Citrus sinensis* | | Lliso et al., 2007 |
| *Pinus nigra* | Australian pine | Wang et al., 2006 |
| *Pinus radiate* | Californian pine | Fiorani Caledon et al., 2007 |
| *Eucalyptus grandis* | rose gum eucalyptus | Lippert et al., 2007 |
| *Picea sitchenisis* | sitka spruce | Valledor et al., 2008 |
| *Pyrus communis* | conference pears | Pedreschi et al., 2007 |

Table 1. Proteomics analyses perfomed on model and non-model plants

providing knowledge that will facilitate the improvement of crop quality, either in terms of resistance to biotic and abiotic stress, or in terms of nutritional processing quality (Salekdeh & Komatsu, 2007; Finnie et al., 2011).

## 5. Proteomic approaches

Comparative plant proteomic approaches are still largely based on traditional two dimentional polyacrilamide gel electrophorsis (2D PAGE) with isoelectric focusing in the first dimension and SDS-PAGE in the second dimension. This technique was initially considered the most suitable method to visualize the differences between protein samples derived from samples grown under different conditions and/or from different tissues. Complex protein mixtures can be resolved efficiently, and the detection of differences in bands or spot intensities is intuitive. Currently, it is possible to visualize over 10,000 protein spots, corresponding to over 1,000 proteins, on single 2D gels (Görg et al., 2009). In many cases, however, individual spots may consist of more than one protein. The differences in spot composition can only be identified by means of mass-spectrometry. The quantitative mass-spectrometry-based proteomics field is constantly evolving, with continuous improvement in protocols, machines and software. Most of the early developments quantitative mass-spectrometry-based proteomic applications were driven by research on yeast and mammalian cell lines. However, in plant physiology analyses, mass spectrometry-based proteomics is no longer only used as a descriptive tool. Instead, well-designed quantitative proteomics has been applied to various aspects of organelle biology, growth regulation and signaling (Schulze & Usadel, 2010). These efforts have greatly improved our knowledge of protein diversity during complex processes. Encouraging pioneer studies on specific subproteomes in plants have revealed candidate proteins that are phosphorylated under specific stress conditions (Oda et al., 1999; Benshop et al., 2007; Niittylä et al., 2007) or during the light independent cycle of photosynthesis (Reiland et al., 2009). Protein abundance changes have been monitored in response to heat shock (Palmblad et al., 2008), during leaf senescence (Hebeler et al., 2008) and during the protein turnover of photosynthetic proteins, monitored using pulse-chase labeling in combination with mass-spectrometry (Nowaczyk et al., 2006). The combination of subcellular fractionation techniques and mass-spectrometry has led to the extensive characterization of the plant subcellular proteome which in turn has led to the discovery of new metabolic pathways (Dunkley et al., 2006). Organelle proteomes were also characterized, such as chloroplasts (Kleffmann et al., 2007; Mejaran et al., 2005; Peltier et al., 2000; Pevzner et al., 2001; Reiland et al., 2009) and plasma membranes and their microdomains (Kierszniowska et al., 2009; Nelson et al., 2006a).

## 6. Problems with proteomic analyses

Although quantitative methods and their results are desirable, the proteomics data that is usually produced is very complex and often variable in quality. The main problem is incomplete data, since the most advanced mass spectrometers cannot sample and fragment all the peptide ions that are present in complex samples. In fact, only a subset of the peptides and proteins present in a sample can be identified. The first step in primary data extraction is the manual validation of the identity of a peptide and quantification through the revision of the spectra assigned to each sequence. The identification of proteins through the use of algorithms has long been practiced and has been well documented (Eng et al., 1994; Pevzner et al., 2001; Craig & Beavis, 2004; Geer et al., 2004; Tanner et al., 2005). The development of robust algorithms to extract quantitative information from multidimensional proteomic experiments, based on mass spectrometry, is instead a more recent development (Schulze & Usadel, 2010 and references therein). Parallel investigations that provide complete genome sequences for several important agricultural crops will make proteomics-based analyses more useful and increase confidence in proteomic identification and characterization. Unfortunately, genome sequencing is still a relatively new approach and is still fairly expensive therefore most plant species of interest have not yet been sequenced, with consequent gaps in the databases. In such cases, it is possible to exploit the homology-driven proteomics for the characterization of proteomes (Junqueira et al., 2008). The availability of fairly large databases of genomic data from model systems has made it feasible to explore the proteomics of single cell types isolated from complex tissues through a procedure known as laser microdissection. The remainder of this chapter is focused on the use of laser microdissection-assisted proteomic analyses on plant tissues.

## 7. Laser microdissection in plant biology

Plants are considered to have about 40 different cell types (Martin et al., 2001). Therefore, the gene expression profiles, protein levels and chemical composition of these cell types are destined to be different, even when they are directly adjacent to each other. For this reason, it is important that the sampling and analysis of data are generated in an ever more spatio-temporal cognizant manner, to allow for a far greater resolution in gene expression (Moco et al., 2009). For many years, *in situ* hybridization and experiments with transgenic plants expressing promoter-gene reporter fusion constructs have been used to identify the expression of individual genes in specific cell types (Jefferson et al. 1987; reviewed in Balestrini & Bonfante, 2008). While these techniques cannot be developed with a high-throughput capability, there is a clear need to analyze a transcriptome and proteome at the specific cell-type level (Klink et al., 2007, 2009, 2010a, 2010b, 2011a, 2011b). It is well known that cell-type specific differences occur in gene expression. Identifying these differences in gene expression is complicated by the complexity of the cells that compose the tissues and organs. Thus, the primary reason for obtaining gene expression information from specific cell types is to minimize the dilution effect caused by the cellular complexity found in tissues and organs. This limitation has been overcome by the laser microdissection (LM) technique which was first described by Isenberg et al. (1976) and then developed at the NIH (National Institute of Health, U.S.) for the dissection of cells from histological tissue sections (Emmert-Buck et al., 1996). Laser microdissection permits the rapid procurement of selected cell populations from a section of heterogeneous tissue in a manner conducive to the extraction of DNA, RNA or proteins. Since it was re-designed for histological sections, LM technology has been used routinely in mammalian (Kamme et al., 2003; Kim et al., 2003; Mouledous et al., 2003) and, in more recent years, in plant systems (Asano et al., 2002; Nakazano et al., 2003; Kerk et al., 2003; Day et al., 2005; Klink et al., 2005). The LM apparatus is generally attached to a light microscope and the dissection of the region of interest is computer-controlled. Several instruments are commercially available to isolate individual cells or groups of cells from intact tissues and they are based on two major methods: laser capture microdissection (LCM) and laser cutting (Day et al., 2005; Nelson et al., 2006b). In LCM, the target cells are attached to a thermoplastic film, which covers an optically clear tube cap, using a pulsed infrared laser. The laser is manipulated so that it melts and fuses the film onto the desired cells. When the cap is removed, the target region is selectively pulled away from the surrounding tissues (Emmert-Buck et al., 1996). An alternative approach uses a UV laser to excise target regions from tissue sections. In the first system, the excised fragment is catapulted upwards into a tube cap (laser microdissection pressure catapulting, LMPC), whereas in the second, the sample falls into the collection tube without any extra forces (LMD). These two instruments allow the collection of a single cell and/or a group of cells or tissue regions. A new generation of LCM systems includes both an infrared laser and a UV laser that allow both laser excised microdissection and capture. Some recent reviews have highlighted the increasing interest of the scientific community in the application of this approach in plant biology (Day et al., 2005, 2006; Nelson et al., 2006b; Ramsay et al., 2006; Balestrini et al., 2009). The preparation of plant samples has been described extensively (Asano et al. 2002; Nakazono et al. 2003; Kerk et al., 2003; Inada & Wildermuth 2005; Klink et al. 2005; Tang et al., 2006; Yu et al., 2007; Balestrini et al., 2007; Klink et al. 2007) with additional details being provided in several reviews (Day et al., 2005, 2006; Nelson et al., 2006b).

## 8. Tissue processing for LM

The tissues for LM are first fixed and sectioned and then the target cells are isolated from the non-target cells under the LM microscope. Sample preparation for LM requires a balance between two contrasting aims: to preserve enough visual detail to identify specific cells during the harvest, and to allow the maximum subsequent recovery of the nucleic acids/proteins from the harvested cells (Figure 1). Two methods have been adopted to prepare sample sections for LM: cryosectioning and paraffin sectioning. Cryosectioning is commonly used in animal research, due to its speed, and it is better at preserving intact molecules, including RNAs and proteins. Although cryosectioning has been described in plant studies (Nakazono et al., 2003), its applicability should be judged on a case-by-case basis (V.K., unpublished observations). Freezing procedures can cause the formation of ice crystals inside vacuoles and air spaces between cells in mature plant tissues: both these features compromise tissue cytology, and eventually lead to the disassembly of cell structures. Cryosectioning of more mature or vacuolated plant material generally requires fixation as well as a cryoprotectant treatment using for example 10–15% sucrose, in order to alleviate the tissue damage caused by freezing. As an alternative, samples are embedded in paraffin after fixation when a more satisfactory preservation of tissue histology is required for target identification. Although this protocol provides excellent cytology, the RNA and protein yield is reduced compared with that from frozen samples. Therefore, it is clear that tissue fixation and paraffin embedding could result in a considerable loss in quality and quantity of the extracted material during RNA studies (Ramsay et al., 2006). Nevertheless, satisfactory amounts of RNA have been obtained from paraffin-embedded material (Kerk et al., 2003; Klink et al., 2005; Tang et al., 2006; Klink et al., 2007, 2009; Hacquard et al., 2010) and an improved morphology is sometimes essential to identify the appropriate cell types for collection purposes. The embedding of *Medicago truncatula* roots in Steedman's wax has recently been used as an alternative to paraffin, and sections of satisfactory morphology and improved RNA quality have been obtained (Gomez & Harrison, 2009). A method for preparing serial sections that reduces RNA degradation has been recently described by using a microwave method (Takahashi et al., 2010). As far as the analysis of nucleic acids is

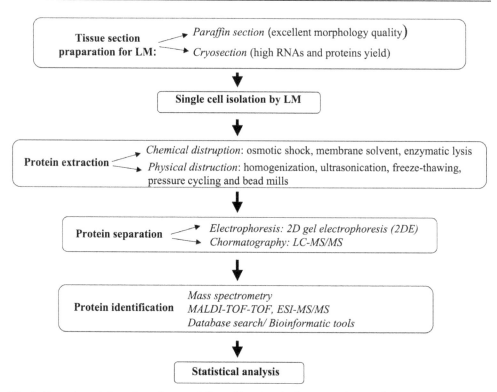

Fig. 1. Experimental proteomics workflow. The classical proteomics workflow has been adapted for a targeted analysis of microdissected samples.

concerned, the possibility of amplifying the RNA extracted from laser microdissected cells allows a transcriptome to be explored by means of microarrays (Nakazono et al., 2003, Casson et al., 2005; Jiang et al., 2006; Klink et al., 2007, 2009; Hacquard et al., 2010) or mRNA-seq techniques based on pyrosequencing platforms, such as 454 Roche and Illumina/Solexa (Graveley, 2008; Simon et al., 2009).

In recent years, LM technology has been applied to gene expression analysis on specific plant cell-types (Day et al., 2005; Nelson et al., 2006b; Ohtsu et al., 2007; Balestrini & Bonfante, 2008; Day et al., 2006; Nelson et al., 2008). The gene expression profile of a number of plant vegetative tissues or cell types, including root cortical cells, vascular bundles, parenchyma, meristem, incipient leaves, syncytia developed from nematode parasitism and abscission zones have been analyzed using the LM technique in several plants (Klink et al., 2005; 2007, 2009, 2010a, 2010b, 2011a, 2011b; Ramsay et al., 2006; Cai & Lashbrook, 2008; Augusti et al., 2009; Nelson et al., 2008 and reference therein). Recently, LM has also been used to provide new insight into fruit development and physiology through the collection of epidermal and subepidermal cells from green, expanding *Citrus clementina* fruit (Matas et al., 2010). A few studies have also focused on the application of LM to gene expression in plant-microbe interactions (Tang et al., 2006; Balestrini et al., 2007; Gomez et al., 2009; Guether et al., 2009a, 2009b; Fiorilli et al., 2009; Chandran et al., 2010; Hacquard et al., 2010).

## 9. Proteomics/metabolomics and LM

The proteome varies in different cells and various cells respond differently to physiological perturbations. Obtaining a better understanding of tissue complexity could be accomplished by isolating specific cells and analyzing them through proteomic analyses, that could compliment mRNA studies. Over the last few years, the combined use of LM and proteomic analysis has been widely adopted in animal biology and significant progress has been made in adapting the technology to the study of plant cellular processes (Gutstein & Morris, 2007). A list of papers on the application of LM in proteomic and metabolomic studies in plant biology is showed in Table 2. However, difficulties in upstream tissue processing, for example achieving cellular morphological integrity and extracting specific types of protein from cells have limited the efficiency of this approach. The most critical step involves extracting as many proteins as possible from the sample of interest. The wide range of chemical properties of proteins implies that the extraction of all the different types of proteins cannot occur with the same efficiency. Despite these difficulties, recent studies have shown that it is possible to obtain useful information from samples as small as those of single cells (Rubakhin et al., 2003; Hummon et al., 2006). Two general classes of fixatives are usually used in LM analysis: cross-linking and precipitating. Cross-linking fixatives generally have little effect on genomic DNA recovery, but have profound effects on RNA (Goldsworthy et al., 1999) and proteins (Rekhter et al., 2001). Therefore, precipitating fixatives such as ethanol and Methacarn are preferred for protein work (Shibutani et al., 2000; Ahram et al., 2003). It has been demonstrated that brief ethanol post fixation and LM using the IR-laser method does not adversely affect proteomic profiling by 2DE (Banks et al., 1999). In plant biology UV laser seems the most used for proteomic studies (Table 2). This could be probably related to the fact that in more recent years the UV-laser systems are the more widespread and also instruments with IR laser cell capture are combined with UV-laser tissue cutting (Balestrini et al., 2009; Nelson et al., 2006b). It has also been showed that paraffin embedding can have only a slight effect on proteomic profiling whether the tissue is processed properly (Ahram et al., 2003; Hood et al., 2006). This is an interesting observation because it opens the way towards the proteomics analyses of LM-collected cells, above all for plant tissues that are particularly prone to cell morphology damage during cryosectioning. Several studies on animal systems have suggested the staining of the tissue section with such dyes as hematoxylin and eosin to guide the dissection process. However, it has been demonstrated that conventional histological staining methods such as cresyl, hematoxylin/eosin and tolouidine blue, as well as some non-conventional methods such as chlorazol black E and Sudan black B, are incompatible with the 2DE-based proteomic analysis of samples isolated by LM (Banks et al., 1999; Craven & Banks, 2001; Moulédous et al., 2002; Craven et al., 2002; Sitek et al., 2005).

As previously mentioned, many efforts have been made to ensure that sample collection methods involving LM do not interfere with the subsequent proteomic analysis. Extractions can be performed both physically and chemically, or as a combination of mechanical disruption and chemical treatments. A wide range of methods has been described to physically disrupt cells for protein analysis: homogenization, ultrasonication, freeze-thawing, pressure cycling, and bead mills (Butt & Coorssen, 2006; Rabilloud et al., 1996). Cellular homogenization and ultrasonication methods are generally more applicable for a wide variety of biological samples. Chemical extraction and protein solubilization have improved substantially over the past few years. The used approaches include denaturation,

| Subject | Tissue preparation | LM system | Technique | Reference |
|---|---|---|---|---|
| Optimization of several tissue fixing and embedding procedures, and of protein extraction methods from *Arabidopsis thaliana* stem microdissected vascular bundle | Fixation in - 70% ethanol - ethanol/acetic acid (75:25 v/v) Paraffin embedding (30 μm) Cryosectioning (30 μm) | LMPC (UV) | 2-DE LC-MS/MS | Schad et al., 2005a |
| Comparison of gene expression and protein accumulation in pericycle cells of maize root | Fixation in ethanol/acetic acid 3:1 Cryosectioning (10 μm) | PixCell II LCM | 2-DE ESI-MS/MS | Dembinsky et al., 2007 |
| Analysis of tissue-specific differences in proteome profiles during barley grain development | Cryosectioning (20 μm) | LMPC (UV) | nanoUPLC combined with ESI-Q-TOF MS | Kaspar et al., 2010 |
| Micromethod for the analysis of amino acid concentrations in NP and ETC cell-type populations from developed barley grain | Cryosectioning (15 μm) | LMPC (UV) | UPLC | Thiel et al., 2009 |
| Metabolite measurement in microdissected vascular bundle samples from *A. thaliana* stem | Cryosectioning (30 μm) | LMPC (UV) | GC-TOF MS | Schad et al., 2005b |
| Analysis of cell wall carbohydrates from lignified and unlignified parenchyma cells, and xylem fibres of *Urtica dioica* | Fixation in 0.2% glutaraldehyde and 2% formaldehyde Paraffin embedding (4 μm) | LCM (UV) | GC-MS | Angeles et al., 2006 |
| Identification of secondary plant metabolities in specific cells from Norway spruce | Cryosectioning (30 μm) | LMD (UV) | NMR MS | Li et al., 2007 |

| Subject | Tissue preparation | LM system | Technique | Reference |
|---------|-------------------|-----------|-----------|-----------|
| Analysis of metabolite profiling in leaf and flower secretory cavities from fresh and dried sample of *Dilatris* plants | Cryosectioning (60 μm) | LMD (UV) LMPC (UV) | NMR HPLC | Schneider & Hölscher, 2007 |
| Combined analysis of RNA transcripts abundance, enzyme activity and metabolite profiles in individual specialized tissues from white spruce stems | Cryosectioning (25 μm) | LMD (UV) | GC-MS | Abbot et al., 2010 |

Table 2. Application of LM in proteomic and metabolomic studies in plant biology

osmotic shock, the use of membrane solvents and enzymatic lysis (Asenjo & Andrews, 1990; Hopkins et al., 1991). When using chemical methods, it is important to reduce the interactions between the proteins, as well as the interactions between the proteins and other substances, including nucleic acids and lipids. It is also important to remove contaminants and interfering substances, and prevent protein precipitation during the separation process (Rabilloud et al., 1996, Gutstein & Morris, 2007). Once the proteins have been extracted, the resultant complex mixture needs to be separated for the subsequent detection, abundance and differential expression analyses.

## 10. Separation technologies used for proteins isolated from LM cells

One of most common methods used to perform protein quantification, which can be coupled with LM technology, is 2D gel electrophoresis (Table 2). At the same time, advances in high-efficiency liquid chromatography (LC), in conjunction with tandem mass spectrometry (MS/MS) have also been reported (Table 2). Although the application of LM to plant biology has been focused above all on cell-specific gene expression profiling, its application to protein analysis has rarely been reported for plant tissues (Nelson et al., 2008; Balestrini & Bonfante, 2008; Hölscher & Schneider, 2008).

This is probably because of the difficulties encountered due to the relatively large amount of proteins that are needed to achieve successful protein profiling (Schad et al., 2005a). As previously mentioned, unlike transcript profiling, which can be performed from very small sample amounts due to efficient amplification strategies, no *in vitro* amplification procedure is yet available for proteins. However, the applicability of 2-DE and high-efficiency liquid chromatography (LC), in conjunction with tandem mass spectrometry (MS/MS), to plant LM material has recently been demonstrated (Schad et al., 2005a). Schad and colleagues (2005a) have compared and optimized several tissue fixation and embedding procedures to obtain the cross sections of *Arabidopsis thaliana* stem tissue, which enabled the microdissetion of vascular bundles, as well as an efficient extraction of proteins. They demonstrated that cryosectioning retains a reasonable morphology and, at the same time, allows an efficient protein extraction. The analysis of proteins from 5000 vascular bundles (~ 250,000 cells yielding about 25 µg total protein) by means of analytical 2-DE has indicated that this tissue processing procedure does not lead to protein degradation/modification. Furthermore, they also optimized the LC-MS/MS approach, starting from a lower amount of material (400 vascular bundles, ~ 20,000 cells, about 2 µg total protein). This resulted in the identification of 131 proteins from 20 stem sections without vascular bundles and 33 specific proteins from 400 vascular bundles. The advantages of the LC-MS/MS approach include the possibility to use a lower amount of material, the capacity for high throughput, no bias against protein classes and high detection ability. The work of Schad et al. (2005a) has certainly increased interest in the application of this procedure, demonstrating that it is a very promising alternative for tissue-specific protein profiling. The number of studies that have employed LM techniques for protein identification and profiling in plant cells has increased significantly over the last years. For example, Dembinsky and colleagues (2007) have analyzed the transcriptome and proteome of pericycle cells in the primary root of maize (*Z. mays*) *versus* non-pericycle cells. For the proteomics experiments, about 1,000 rings of pericycle cells (200,000 cells) have been isolated from root cross sections, extracting approximately 30 µg of proteins, which were separated by 2-DE. The 56 most abundant protein spots were picked from a representative 2-D gel, digested with trypsin and the eluted peptides were subjected to liquid chromatography-tandem mass spectrometry (LC-MS/MS). The pericycle reference map was made in triplicate from indipendent protein preparations and all the identified proteins were detected in all the replicates. Twenty of the 56 proteins were identified by matching known plant proteins, thus defining a reference dataset of the maize pericycle proteome. In another study, Kaspar et al. (2010) focused their attention on tissue-specific differences in the proteome during barley grain development. In order to address this issue, nucellar projection (NP) and endosperm transfer cells (ETC) of barley grain were collected by LM. Proteins were subsequently extracted, digested with trypsin and analyzed through nanoLC separation combined with ESI-Q-TOF mass spectrometry. This procedure requires material from between 40 and 75 sections per sample. Three independent extractions showed highly reproducible chromatograms. Quantitative and qualitative protein profiling led to the identification of a number of proteins with tissue specific expression. For example, 137 proteins were identified from ETC and 44 from the NP. Among the identified proteins, 31 were identified in both tissues. The major differences between ETC and NP protein profiles concerned cell wall and protein synthesis (in the ETC but not in the NP) and the disease response (with a greater representation in NP), which is in agreement with previously published transcript analyses (Thiel et al., 2008). These experiments have shown that nanoLC-based separation in combination with MS detection can be considered a suitable platform for identifying proteins present in laser-microdissected samples, which contains only small quantities of proteins (Kaspar et al., 2010).

## 11. Metabolomic studies in cells isolated by LM

The last decade has seen an increase in metabolomic-based studies, which are crucial to understand cellular processes because they can connect metabolite profiles and metabolic changes to protein activity, and thus leading to a detailed and more comprehensive understanding of the phenotype of the organism of interest. So far, studies in this field have mainly been performed on whole plants, organs, such as fruits (Moco et al., 2006; Fraser et al., 2007), leaves (Kant et al., 2004; Glauser et al., 2008), tubers (Roessner et al., 2001; Sturm et al., 2007), flowers (Kazuma et al., 2003; Wang et al., 2004), and roots (Opitz et al., 2002; Hagel et al., 2008;). However, some studies have also been performed on specific tissues (Moco et al., 2007; Fait et al., 2008) and even on specific cells (Li et al., 2007; Schneider & Hölscher, 2007). Metabolite analysis at a microscale level from sectioned tissues or cells is a major challenge since metabolities (usually < 1500 Da) show an enormous chemical diversity and for this reason general multiple approaches are required for extraction, fractionation and analysis. Moreover, there is a higher turnover of metabolites than large biomolecules and there is a dynamic range of metabolite concentrations. Micromethods have been adapted from animal biology in order to determine the spatial distribution of small molecules in plant tissue (Schneider & Hölscher 2007; Fait et al., 2008; Hagel et al., 2008). Among the two different methods of LM, laser capture microdissection (LCM) and laser cutting, this last seems to be the most useful method for harvesting samples for metabolite analysis because, in contrast to LCM, it is contact-free and avoids potential contamination from the melting foil (Moco et al., 2009). In addition, most of the analyses have exploited the cryosection method, thus avoiding any further chemical treatment of the material (Table 2). Using standard tissue fixation and embedding protocols, metabolites can in fact either be extracted by means of dehydrating solvents, or washed out by embedding agents (Schad et al., 2005b). Paraffin embedding has been used for the carbohydrate analysis of the polysaccharides from the walls of lignified and unlignified parenchyma cells, and of xylem fibres of *Urtica dioica* (Angeles et al., 2006). The carbohydrate composition of different cell wall types was obtained by the combination of laser microdissection and GC-MS analysis.

For metabolite analyses based on LM, GC-TOF-MS, LC-MS, GC-MS and NMR-related strategies have been used (Schad et al., 2005b; Lisec et al., 2006; Moco et al., 2006). MS-based analytical methods probably ensure a higher identification power for small molecules than NMR measurements. In the first study in which LM was applied successfully to analyze the spatial distribution of metabolites in plant tissues, Schad and colleagues (2005b) used the GC-TOF MS technique to investigate vascular bundles obtained from *Arabidopsis thaliana* cross sections. Cryo-sectioned stem material of 30 μm section thickness was subjected to LMPC. Vascular bundles were dissected and catapulted into the collection device, which was filled with ethanol to inactivate the metabolic enzymes and protect the cell contents from undesired enzymatic modification. An ethanol extract of approximately 100 collected vascular bundles (~5,000 cells) was derivatized with N-methyl-N-(trimethylsilyl) trifluoroacetamide (MSTFA) and subjected to GC-time-of-flight (TOF) MS analysis to simultaneously detect compounds of different classes. Sixty-eight metabolites were detected in the vascular bundles; sixty-five metabolites were instead identified in control samples, which are sections without vascular bundles.

As an alternative, Thiel et al. (2009) used a combination of LMPC-based microdissection and liquid chromatography (UPLC) to analyze the amino acid concentrations in nucellar projections (NP) and endosperm transfer cells (ETC) from developing barley grains. In order to guarantee a sufficient amount of material to produce consistent values and detect the differences in the amino acid concentrations between the two tissues, the authors prepared 10-20 cryosections for one sample and analyzed 4-5 biological replicates/sample. UPLC technology was used to measure free amino acid concentrations from microdissected tissues and the sum of all the measured amino acids was 98 and 112 amol m$\mu^{-3}$ for NP and ETC, respectively. This metabolite approach based on LM was combined with a transcriptome analysis. On the basis of these studies, it has been concluded that combining metabolite data with a transcriptome approach leads to a better understanding of the metabolism, interconversion and transfer of amino acids at the maternal–filial boundary of growing barley seeds.

Methods have been also developed to analyze laser-microdissected samples by means NMR spectroscopy (http://www.ice.mpg.de/ext/769.html). For instance, high-resolution [1]H NMR spectroscopy has been used, in combination with LM, as a tool to analyze the contents of the secretory cavities from fresh leaves and herbarium specimen of *Dilatris* plants (Haemodoraceae) (Schneider & Hölscher, 2007). The secretory cavity sections show a typical storage cell surrounded by a thin layer of glandular epithelial cells. Their low water content makes them well accessible to LM (Moco et al., 2009). The dissected cavities were localized under a stereomicroscope. They were then picked up using an extremely sharp dissecting needle and transferred directly to a microcentrifuge tube containing the extraction solvent (acetone/water 20:1). In some experiments, the dissected material was transferred directly to the NMR tube without centrifugation, and extracted using the NMR solvent (deuterated acetone) in an ultrasonic bath. The extracts were subjected to cryogenic 1H NMR spectroscopy and reversed-phase high-performance liquid chromatography (HPLC). The results obtained from 180-year-old herbarium specimens of *Dilatris corymbosa* and *Dilatris viscosa* showed that phenylphenalenones, which are typical secondary metabolites of Hemodoraceae, were identified in secretory cells of leaves and flower petals (Schneider & Hölscher, 2007).

LM has not been widely applied to woody plant tissue. Cell-specific metabolic profilings have been conducted on special cells harvested from the bark of Norvegian spruce (*Picea abies*) (Li et al., 2007) by means a combination of LM, NMR, and MS. Sclereids (stone cells) were detected in cryosections of the bark taking advantage of their characteristic fluorescence and this was followed by laser microdissection. Non-fluorescing phloem tissue was microdissected from the same cryosections and used as a control sample. The collected samples were then transferred to NMR tubes to which deuterated methanol was added for extraction. [1]H and 2D NMR spectra were measured using a cryogenically cooled probehead. The results indicate that both sclereids and the adjacent parenchymatic tissue show similar phenolic components. Comparison with the spectra of reference compounds, together with MS analysis, revealed that astringin (major component) and dihydroxyquercetin 3'-O-β-**D**-glucopyranoside (minor component) are present in both the sclereids and the control cells. The control cells (sclereid-surrounding cells) showed higher levels of the two components.

Abbott and colleagues (2010) have recently reported, in a methodology article, the successful use of LMD technology to isolate individual specialized tissues from the stems of the woody perennial *Picea glauca* (white spruce), suitable for subsequent combined analysis of RNA transcripts abundance, enzyme activity and metabolite profiles. In agreement with previous papers, the authors underlined that sample preparation protocols for LM can vary substantially on the basis on the type of tissue and down-stream analysis. A tangential cryosectioning approach was essential to obtain large quantities of cortical resin ducts (CRD) and cambial zone (CZ) tissues using LM. Gene expression results showed a differential expression of genes involved in terpenoid metabolism between the CRD and CZ tissues, and in response to methyl jasmonate (MeJA). In addition, terpene synthase enzyme activity has been identified in CZ protein extracts and terpenoid metabolites were detected, by means of GC-MS, in both the CRD and CZ tissues. These analyses supported by LM seem to be very promising to improve the characterization of complex processes related to woody plant development, including cell differentiation and specialization associated with stem growth, wood development and the formation of defense-related structures such as resin ducts.

## 12. Bioinformatics

In 2002, Scheidler and colleagues demonstrated altered gene activity in *Arabidopsis* infected with *Phytopthora*. The work provided a meaningful context for the gene expression analyses that were performed, and resulted in the identification of the major shifts in physiology and metabolism that occur during the infection process. However, the analyses focused on gene expression in whole infected plants. Unlike Scheideler et al. (2002), Klink et al. (2011b) and

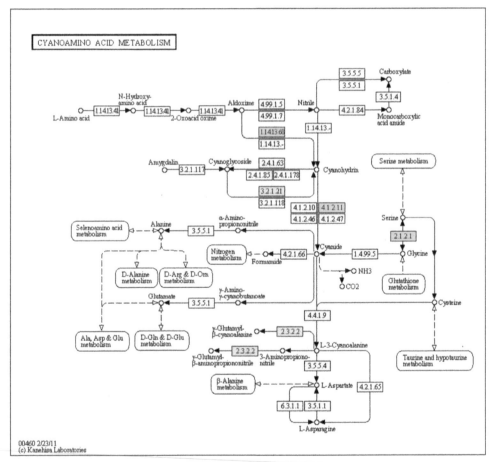

Fig. 2. A PAICE pathway for cyanoamino acid metabolism for 3 day post infection syncytia undergoing a resistant reaction in *G. max* as it is being infected by *Heterodera glycines* (soybean cyst nematode). The green boxes represents active genes (adapted from Matsye et al., submitted).

Matsye et al. (submitted) have used the same principle, adapting the publically available Kyoto Encyclopedia of Genes and Genomes (KEGG) (http://www.genome.jp/kegg/catalog/org_list.html) and modifying it so that gene expression can be visualized using a KEGG application called Pathway Analysis and Integrated Coloring of Experiments (PAICE). PAICE was developed in the laboratories of Dr. Benjamin Matthews (USDA; Beltsville, MD) and Dr. Nadim Alkharouf (Towson University, Baltimore, MD) (Hosseini et al., unpublished) and is freely available (http://sourceforge.net/projects/paice/). PAICE has been used on LM cells infected with parasitic nematodes, and it provides a deeper understanding of the biochemical and metabolic activities during multiple defense reactions in multiple G. *max* genotypes compared to both pericycle control cell populations and the susceptible reaction (Klink et al., 2011; Matsye et al., submitted). However, the analyses were based on RNA isolated from the specific cell types and not on proteins or metabolites (Figure 2). It is believed that PAICE could be expanded to provide a comprehensive understanding of any cell isolated by LM and analyzed for its proteomic and metabolic content.

## 13. Conclusion

To have a better understanding of tissue and organ-defined processes and functions, it is necessary to study the biochemical activity at a cellular resolution level by analyzing the proteome. This has become increasingly important, since it has been demonstrated, in several comparative studies, that protein expression and abundance often poorly correlate with the mRNA levels in the same cell types (Schad et al., 2009a). Many proteins are the primary determinant molecules of physiological processes and are often restricted to specific tissues and cell types. Thus, the monitoring of protein expression at a very high spatial resolution could help enhance our understanding of the biological processes that control plant growth and development. At the same time, the use of different strategies and protocols for the characterization of a wide number of metabolites from a single cell or tissue have increased significantly over the last decade making the broad applicability of these analyses tractable. In order to address these issues, sampling methods, for example LM in plants, have been adopted to extract highly specific tissue regions and homogeneous cell-type populations with limited damage, and have led to the discovery of functions of genes/proteins/metabolites that contribute to cell specialization (Galbraith & Birnbaum 2006). Despite these considerable efforts, the current strategies used for protein/metabolite characterization still face significant obstacles. These challenges are mainly caused by the cellular complexity and spatial and temporal distribution of localized gene activity within living tissues, including metabolic processes. Other challenges concern the identification of the high degree of chemical diversity of the different cell types that can be affected by the analysis procedures. Technical improvements are still required to achieve reliable protein and metabolite profilings in small samples. The introduction of statistical analysis, applied to the handling and manipulation of data from proteomics and metabolomics, will lead to the development of promising strategies that can be used to extract precious information from large data sets and to identify new proteins and metabolites. Although most of these restrictions have already been solved in the field of genomics and transcriptomics, the problem still remains of adapting these computational strategies for proteomic and metabolic analyses.

## 14. Acknowledgements

Contributions to this chapter have been partly funded by CNR (Premio DAA 2009) to RB. VF was supported by a grant from BIOBIT-CIPE (Piedmont Region project). VPK would like to thank the Mississippi Soybean Promotion Board for the funding and critical reading of the manuscript by Prachi D. Matsye.

## 15. References

Abbott, E., Hall, D., Hamberger, B. & Bohlmann, J. (2010). Laser microdissection of conifer stem tissues: isolation and analysis of high quality RNA, terpene synthase enzyme activity and terpenoid metabolites from resin ducts and cambial zone tissue of white spruce (*Picea glauca*), *BMC Plant Biology* 10:106.

Ahram, M., Flaig, M.J., Gillespie, J.W., Duray, P.H., Linehan, W.M., Ornstein, D.K., Niu, S., Zhao, Y., Petricoin, E.F. 3rd & Emmert-Buck, M.R. (2003). Evaluation of ethanol-fixed, paraffin-embedded tissues for proteomic applications, *Proteomics* 3: 413–421.

Ali, G.S. & Reddy, A.S.N. (2008) Regulation of Alternative Splicing of Pre-mRNAs by Stresses, *Current Topics in Microbiology and Immunology* 326: 257-275.

Agrawal, G.K. & Rakwal, R. (2011). Rice proteomics: A move toward expanded proteome coverage to comparative and functional proteomics uncovers the mysteries of rice and plant biology, *Proteomics* 11: 1630–1649.

Angeles, G., Berrio-Sierra, J., Joseleau, J.P., Lorimier, P., Lefebvre, A. & Ruel, K. (2006). Preparative laser capture microdissection and single- pot cell wall material preparation: a novel method for tissue-specific analysis, *Planta* 224: 228–232.

Asano, T., Masumura, T., Kusano, H., Kikuchi, S., Kurita, A., Shimada, H. & Kadowaki, K. (2002). Construction of a specialized cDNA library from plant cells isolated by laser capture microdissection: toward comprehensive analysis of the genes expressed in the rice phloem, *The Plant Journal* 32: 401-408.

Asenjo, J.A. & Andrews, B.A. (1990) Enzymatic Cell Lysis for Product Release, in J.A. Asenjo, Marcel Dekker (eds), *Separation Processes in Biotechnology*, New York, pp. 143-175.

Augusti, J., Merelo, P., Cercós, M., Tadeo, F.R. & Talón, M. (2009) Comparative transcriptional survey between laser-microdissected cells from laminar abscission zone and petiolar cortical tissue during ethylene-promoted abscission in citrus leaves, *BMC Plant Biology* 9: 127.

Baerenfaller, K., Grossmann, J., Grobei, M.A., Hull, R., Hirsch-Hoffmann, M., Yalovsky, S., et al. (2008). Genome-scale proteomics reveals *Arabidopsis thaliana* gene models and proteome dynamics, *Science* 320: 938–41.

Balestrini, R., Gòmez-Ariza, J., Lanfranco, L. & Bonfante, P. (2007). Laser Microdissection reveals that transcripts for five plant and one fungal phosphate transporter genes are contemporaneously present in arbusculated cells, *Molecular Plant–Microbe Interactions* 20: 1055–1062.

Balestrini, R. & Bonfante, P. (2008). Laser Microdissection (LM): Applications to plant materials, *Plant Biosystems* 142: 331-336.

Balestrini, R., Gòmez-Ariza, J., Klink, V.P. & Bonfante P. (2009). Application of laser microdissection to plant pathogenic and symbiotic interactions, *Journal of Plant Interactions* 4: 81-92.

Banks, R.E., Dunn, M.J., Forbes, M.A., Stanley, A., Pappin, D., Naven, T., Gough, M., Harnden, P. & Selby, P.J. (1999). The potential use of laser capture microdissection to selectively obtain distinct populations of cells for proteomic analysis - Preliminary findings, *Electrophoresis* 20: 689–700.

Barrett, T.H. & Gould H. (1973) Tissue and species specificity of non-histone chromatin proteins, *Biochimica et Biophysica Acta* 294: 165-170.

Benschop, J.J., Mohammed, S., O'Flaherty, M., Heck, A.J., Slijper, M. & Menke, F.L. (2007). Quantitative phospho-proteomics of early elicitor signalling in *Arabidopsis*, *Molecular and Cellular Proteomics* 6: 1705–13.

Bligny, R. & Douce, R. (2001) NMR and plant metabolism. *Current Opinion in Plant Biology* 4: 191-196.

Brownfield, L., Ford, K., Doblin, M.S., Newbigin, E., Read, S. & Bacic, A. (2007). Proteomic and biochemical evidence links the callose synthase in *Nicotiana alata* pollen tubes to the product of the *NaGSL1* gene, *The Plant Journal* 52: 147–56.

Butt, R.H. & Coorssen, J.R. (2006). Pre-extraction Sample Handling by Automated Frozen Disruption Significantly Improves Subsequent Proteomic Analyses, *Journal of Proteome Research* 5: 437-448.

Cai, S. & Lashbrook, C.C. (2006). Laser capture microdissection of plant cells from tape-transferred paraffin sections promotes recovery of structurally intact RNA for global gene profiling, *The Plant Journal* 48: 628–637.

Campbell, M.A., Haas, B.J., Hamilton, J.P., Mount, S.M. & Buell, C.R. (2006). Comprehensive analysis of alternative splicing in rice and comparative analyses with *Arabidopsis*, *BMC Genomics* 7:327.

Casson, S., Spencer, M., Walker, K. & Lindsey, K. (2005). Laser capture microdissection for the analysis of gene expression during embryogenesis of *Arabidopsis*, *The Plant Journal* 42: 111-123.

Celedon, P.A.F., Andrade, A., Meireles, K.G.X., Carvalho, M.C.C.G., Caldas, D.G.G., Moon, D.H., Carneiro, R.T., Franceschini, L.M., Oda, S. & Labate, C.A. (2007). Proteomic analysis of the cambial region in juvenile *Eucalyptus grandis* at three ages, *Proteomics* 7: 2258-2274.

Chandran, D., Inada, N., Hather, G., Klindt, C.K. & Wildermuth M.C. (2010). Laser microdissection of *Arabidopsis* cells at the powdery mildew infection site reveals site-specific process and regulators, *Proceedings of the National Academy of Science USA* 107: 460-465.

Chen, B.E., Kondo, M., Garnier, A., Watson, F.L., Püettmann-Holgado, R., Lamar, D.R. & Schmucker, D. (2006). The molecular diversity of Dscam is functionally required for neuronal wiring specificity in *Drosophila*, *Cell* 125: 607-620.

Ciobanu, L. & Pennington, C.H. (2004). 3D micron-scale MRI of single biological cells, *Solid State Nuclear Magnetic Resonance* 25: 138-141.

Craig, R. & Beavis, R.C. (2004). TANDEM: matching proteins with mass spectra, *Bioinformatics* 20: 1466-67.

Cravatt, B.F., Simon, G.M. & Yates, III J.R. (2007). The biological impact of mass-spectrometry-based proteomics, *Nature* 450: 991-1000.

Craven, R. & Banks, R. (2001). Laser capture microdissection and proteomics: Possibilities and limitation, *Proteomics* 1: 1200-1204.

Craven, R.A., Totty, N., Harnden, P., Selby, P.J. & Banks, R.E. (2002). Laser capture microdissection and two-dimensional polyacrylamide gel electrophoresis:

evaluation of tissue preparation and sample limitations, *American Journal of Pathology* 160: 815–822.

Day, R.C., Grossniklaus, U. & Macknight, R.C. (2005). Be more specific! Laser-assisted microdissection of plant cells, *Trends in Plant Science* 10: 397–405.

Day, R.C., McNoe, L.A. & Macknight, R.C. (2006). Transcript analysis of laser microdissected plant cells, Technical Focus, *Physiologia Plantarum* 129: 267–282.

de Jong, F., Mathesius, U., Imin, N., Rolfe, B.G. (2007). A proteome study of the proliferation of cultured *Medicago truncatula* protoplasts, *Proteomics* 7: 722-36.

Dembinsky, D., Woll, K., Saleem, M., Liu, Y., Fu, Y., Borsuk, L.A., Lamkemeyer, T., Fladerer, C., Madlung, J., Barbazuk, B., Nordheim, A., Nettleton, D., Schnable, P.S. & Hochholdinger F. (2007). Transcriptomic and proteomic analyses of pericycle cells of the maize primary root, *Plant Physiology* 145: 575-88.

Djordjevic, M.A., Oakes, M., Li, D.X., Hwang, C.H., Hocart, C.H. & Gresshoff, P.M. (2007). The *glycine max* xylem sap and apoplast proteome, *Journal of Proteome Research* 6: 3771-9.

Dunkley, T.P., Hester, S., Shadforth, I.P., Runions, J., Weimar, T., Hanton, S.L., Griffin, J.L., Bessant, C., Brandizzi, F., Hawes, C., Watson, R.B., Dupree, P. & Lilley, K.S. (2006). Mapping the *Arabidopsis* organelle proteome, *Proceedings of National Academy Science USA* 103: 6518-23.

Edman, P. (1950). Method for determination of the amino acid sequence in peptides, *Acta Chemica Scandinavica* 4: 283-293.

Emmert-Buck, M.R., Bonner, R.F., Smith, P.D., Chuaqui, R.F., Zhuang, Z., Goldstein, S.R., Weiss, R.A. & Liotta, L.A. (1996). Laser capture microdissection, *Science* 274: 998–1001.

Eng, J., McCormack, A.L. & Yates, J.R.I. (1994). An approach to correlate tandem mass spectral data of peptides with amino acid sequences in a protein database, *Journal of the American Society for Mass Spectrometry* 5: 976–89.

Fait, A., Hanhineva, K., Beleggia, R., Dai, N., Rogachev, I., Nikiforova, V.J., Fernie, A.R. & Aharoni, A. (2008). Reconfiguration of the chene and receptacle metabolic networks during strawberry fruit development, *Plant Physiology* 148: 730–750.

Finnie, C., Sultan, A. & Grasser, K.D. (2011). From protein catalogues towards targeted proteomics approaches in cereal grains, *Phytochemistry* 72: 1145-1153.

Fiorilli, V., Catoni, M., Miozzi, L., Novero, M,. Accotto, G.P. & Lanfranco, L. (2009). Global and cell-type gene expression profiles in tomato plants colonized by an arbuscular mycorrhizal fungus, *New Phytologist* 184: 975-987.

Fraser, P.D., Enfissi, E.M.A., Goodfellow, M., Eguchi, T. & Bramley, P.M. (2007). Metabolite profiling of plant carotenoids using the matrix-assisted laser desorption ionization time-of-flight mass spectrometry, *The Plant Journal* 49: 552–564.

Galbraith, D.W. & Birnbaum, K. (2006). Global studies of cell type-specific gene expression in plants, *Annual Review of Plant Biology* 57:451-75.

Gazanchian, A., Hajheidari, M., Sima, N.K. & Salekdeh, G.H. (2007). Proteome response of *Elymus elongatum* to severe water stress and recovery, *Journal of Experimental Botany* 58: 291–300.

Geer, L.Y., Markey, S.P., Kowalak, J.A., Wagner, L., Xu M, Maynard, D.M., Yang, X., Shi, W. & Bryant, S.H.2004. Open mass spectrometry search algorithm, *Journal of Proteome Research* 3: 958–64.

Glauser, G., Guillarme, D., Grata, E., Boccard, J., Thiocone, A., Carrupt, P.A., Veuthey, J.L., Rudaz, S., & Wolfender, J.L. (2008). Optimized liquid chromatography-mass spectrometry approach for the isolation of minor stress biomarkers in plant extracts and their identification by capillary nuclear magnetic resonance, *Journal of Chromatography* 1180: 90–98.

Goldsworthy, S.M., Stockton, P.S., Trempus, C.S., Foley, J.F. & Maronpot, R.R. (1999) Effects of fixation on RNA extraction and amplification from laser capture microdissected tissue, *Molecular Carcinogenesis* 25: 86-91.

Gomez, S.K. & Harrison, M.J. (2009). Laser microdissection and its application to analyze gene expression in the arbuscular mycorrhizal symbiosis. Pest Management Sciences 65: 504-511

Gomez, K.S., Javot, H., Deewatthanawong, P., Torres-Jerez, I., Tang, Y., Blancaflor, B.E., Udvardi, M.K. & Harrison, J. M. (2009). *Medicago Truncatula* and *Glomus Intraradices* gene expression in cortical cells harboring arbuscules in the arbuscular mycorrhizal symbiosis, *BMC Plant Biology* 9:10.

Görg, A., Drews, O., Lück, C., Weiland, F. & Weiss, W. 2-DE with IPGs, *Electrophoresis* 30: S122-32.

Graveley, B.R. (2008) Molecular biology: power sequencing, *Nature* 453: 1197-8.

Guether, M., Balestrini, R., Hannah, M., He, J., Udvardi, M.K. & Bonfante, P. (2009a). Genome-wide reprogramming of regulatory networks, transport, cell wall and membrane biogenesis during arbuscular mycorrhizal symbiosis in *Lotus japonicus*, *New Phytologist* 182: 200–212.

Guether, M., Neuhauser, B., Balestrini, R., Dynowski, M., Ludewig, U. & Bonfante P. (2009b). A mycorrhizal-specific ammonium transporter from Lotus Japonicus acquires nitrogen released by arbuscular mycorrhizal fungi, *Plant Physiology* 105: 73-83.

Gutstein, H.B. & Morris, J.S. (2007). Laser capture sampling and analytical issue in proteomics, *Expert Review of Proteomics* 4: 627-37.

Hacquard, S., Delaruelle, C., Legué, V., Tisserant, E., Kohler, A., Frey, P., Martin, F. & Duplessis S. (2010) Laser capture microdissection of uredinia formed by *Melampsora larici-populina* revealed a transcriptional switch between biotrophy and sporulation, *Molecular Plant-Microbe Interactions* 23: 1275-1286.

Hagel, J.M., Weljie, A.M., Vogel, H.J. & Facchini, P.J. (2008). Quantitative H-1 nuclear magnetic resonance metabolite profiling as a functional genomics platform to investigate alkaloid biosynthesis in opium poppy, *Plant Physiology* 147: 1805-1821.

Hebeler, R., Oeljeklaus, S., Reidegeld, K.A., Eisenacher, M., Stephan, C., Sitek, B., Stühler, K., Meyer, H.E., Sturre, M.J., Dijkwel, P.P. & Warscheid, B. (2008). Study of early leaf senescence in *Arabidopsis thaliana* by quantitative proteomics using reciprocal 14 N/15 N labeling and difference gel electrophoresis, *Molecular and Cellular Proteomics* 7:108–20.

Hinse, C., Sheludko, Y.V., Provenzani, A., Stöckigt, J.H.H. (2001). *In vivo* NMR at 800 MHz to monitor alkaloid metabolism in plant cell cultures without tracer labeling, *Journal of America Chemical Society* 123: 5118-5119.

Hochholdinger, F., Sauer, M., Dembinsky, D., Hoecker, N., Muthreich, N., Saleem, M. & Liu, Y. (2006). Proteomic dissection of plant development, *Proteomics* 6: 4076–4083.

Hölscher, D., Schneider, B. (2008). Application of Laser-Assisted Microdissection for Tissue and Cell-Specific Analysis of RNA, Proteins, and Metabolites, *Progress in Botany* 69: 141-167.

Hood, B.L., Conrads, T.P. & Veenstra, T.D. (2006). Unravelling the proteome of formalin-fixed paraffin-embedded tissue, *Briefings in Functional Genomics and Proteomics* 5: 169–175.

Hopkins, T.R. (1991). Physical and chemical cell disruption for the recovery of intracellular proteins, *Bioprocess Technology* 12: 57–83.

Hummon, A.B., Amare, A. & Sweedler, J.V. (2006). Discovering new invertebrate neuropeptides using mass spectrometry, *Mass Spectrometry Reviews* 25: 77–98.

Inada, N. & Wildemuth, M.C. (2005). Novel tissue preparation method and cell-specific marker for laser microdissection of *Arabidopsis* mature leaf, *Planta* 221: 9–16.

Ingle, R.A., Schmidt, U.G., Farrant, J.M., Thomson, J.A. & Mundree, S.G. (2007). Proteomic analysis of leaf proteins during dehydration of the resurrection plant *Xerophyta viscosa*, *Plant Cell and Environment* 30: 435-46.

Isenberg, G., Bielser, W., Meier-Ruge, W. & Remy, E. (1976) Cell surgery by laser microdissection: a preparative method, *Journal of Microscopy* 107: 19–24.

Jamet, E., Boudart, G., Borderies, G., Charmont, S., Lafitte, C., Rossignol, M., Canut, H. & Pont-Lezica R.F. Isolation of plant cell wall proteins, *Methods Molecular Biology* 425: 187-201.

Jefferson, R.A., Kavanagh, T.A. & Bevan, M.W. (1987). GUS fusions: β-glucuronidase as a sensitive and versatile gene fusion marker in higher plants, *EMBO Journal* 6: 3901-3907.

Jiang, G., Wang, Z., Shang, H., Yang, W., Hu, Z., Phillips, J. Deng, X. (2007). Proteome analysis of leaves from the resurrection plant *Boea hygrometrica* in response to dehydration and rehydration, *Planta* 225: 1405-20.

Jiang, K., Zhang, S., Lee, S., Tsai, G., Kim, K., Huang, H., Chilcott, C., Zhu, T. & Feldman, L.J. (2006). Transcription profile analyses identify genes and pathways central to root cap functions in maize, *Plant Molecular Biology* 60: 343–363.

Jorrin, J.V., Maldonado, A.M. & Castillejo, M.A. (2007). Plant proteome analysis: a 2006 update, *Proteomics* 7: 2947–62.

Jorrín-Novo, J.V., Maldonado, A.M., Echevarría-Zomeño, S., Valledor, L., Castillejo, M.A., Curto, M., Valero, J., Sghaier, B., Donoso, G. & Redondo, I. (2009). Plant proteomics update (2007-2008): Second-generation proteomic techniques, an appropriate experimental design, and data analysis to fulfill MIAPE standards, increase plant proteome coverage and expand biological knowledge, *Journal of Proteomics* 72: 285-314.

Junqueira, M., Spirin, V., Santana Balbuenaa, T., Thomasa, H., Adzhubeib, I., Sunyaevb, S. & Shevchenko, A. (2008). Protein identification pipeline for the homology-driven proteomics, *Journal of Proteomics* 71: 346-356.

Kamme, F., Salunga, R., Yu, J., Tran, D.T., Zhu, J., Luo, L., Bittner, A., Guo, H.Q., Miller, N., Wan, J. & Erlander, M. (2003). Single-cell microarray analysis in hippocampus CA1: demonstration and validation of cellular heterogeneity, *Journal of Neuroscience* 23: 3607-3615.

Kant, M.R., Ament, K., Sabelis, M.W., Haring, M.A. & Schuurink, R.C. (2004). Differential timing of spider mite-induced direct and indirect defenses in tomato plants, *Plant Physiology* 135: 483- 495.

Kaspar, S., Weier, D., Weschke, W., Mock, H.P. & Matros, A. (2010). Protein analysis of laser capture micro-dissected tissues revealed cell-type specific biological functions in developing barley grains, *Analytical and Bioanalytical Chemistry* 398: 2883-93.

Kazuma, K., Noda, N. & Suzuki, M. (2003). Flavonoid composition related to petal color in different lines of *Clitoria ternatea*, *Phytochemistry* 64: 1133-1139.

Kerk, N.M., Ceserani, T., Tausta, S.L., Sussex, I.M. & Nelson, T.M. (2003). Laser capture microdissection of cells from plant tissues, *Plant Physiology* 132: 27–35.

Kierszniowska, S., Seiwert, B. & Schulze, W.X. (2009). Definition of *Arabidopsis* sterol-rich membrane microdomains by differential treatment with methyl-ß-cyclodextrin and quantitative proteomics, *Molecular and Cellular Proteomics* 8: 612–23.

Kim, E., Magen, A. & Ast, G. (2007). Different levels of alternative splicing among eukaryotes, *Nucleic Acids Research* 35: 125–31.

Kim, J.O., Kim, H.N., Hwang, M.H., Shin, H.I., Kim, S.Y., Park, R.W., Park, E.Y., Kim, I.S., van Wijnen, A.J., Stein, J.L., Lian, J.B., Stein, G.S. & Choi, J.Y. (2003). Differential gene expression analysis using paraffin-embedded tissues after laser microdissection, *Journal Cell Biochemistry* 90: 998-1006.

Kleffmann, T., von Zychlinski, A., Russenberger, D., Hirsch-Hoffmann, M., Gehrig, P, Gruissem, W. & Baginsky, S. (2007). Proteome dynamics during plastid differentiation in rice, *Plant Physiology* 143: 912–23.

Klink, V.P., MacDonald, M., Alkharouf, N. & Matthews, B.F. (2005). Laser capture microdissection (LCM) and expression analyses of *Glycine max* (soybean) syncytium containing root regions formed by the plant pathogen *Heterodera glycines* (soybean cyst nematode), *Plant Molecular Biology* 59: 969-983.

Klink, V.P., Overall, C.C., Alkharouf, N., MacDonald, M.H. & Matthews, B.F. (2007). Laser capture microdissection (LCM) and comparative microarray expression analysis of syncytial cells isolated from incompatible and compatible soybean roots infected by soybean cyst nematode (*Heterodera glycines*), *Planta* 226: 1389-1409.

Klink, V.P., Hosseini, P., Matsye, P., Alkharouf, N. & Matthews, B.F. (2009). A gene expression analysis of syncytia laser microdissected from the roots of the *Glycine max* (soybean) genotype PI 548402 (Peking) undergoing a resistant reaction after infection by *Heterodera glycines* (soybean cyst nematode), *Plant Molecular Biology* 71: 525-567.

Klink, V.P., Hosseini, P., Matsye, P., Alkharouf, N. & Matthews, B.F. (2010a). Syncytium gene expression in *Glycine max*[PI 88788] roots undergoing a resistant reaction to the parasitic nematode *Heterodera glycines*, *Plant Physiology and Biochemistry* 48: 176-193.

Klink, V.P., Overall, C.C., Alkharouf, N., MacDonald, M.H. & Matthews, B.F. (2010b). Microarray detection calls as a means to compare transcripts expressed within syncytial cells isolated from incompatible and compatible soybean (*Glycine max*) roots infected by the soybean cyst nematode (*Heterodera glycines*), *Genome* Article ID 491217: 1-30.

Klink, V.P., Matsye, P.D. & Lawrence, G.W. (2011a). Cell-specific studies of soybean resistance to its major pathogen, the soybean cyst nematode as revealed by laser capture microdissection, gene pathway analyses and functional studies in Aleksandra Sudaric (ed.) *Soybean - Molecular Aspects of Breeding*, Intech Publishers pp. 397-428.

Klink, V.P., Hosseini, P., Matsye, P.D., Alkharouf, N. & Matthews, B.F. (2011b). Differences in gene expression amplitude overlie a conserved transcriptomic program occurring between the rapid and potent localized resistant reaction at the syncytium of the *Glycine max* genotype Peking (PI 548402) as compared to the prolonged and potent resistant reaction of PI 88788, *Plant Molecular Biology* 75: 141-165.

Komatsu, S., Konishi, H. & Hashimoto, M. (2007). The Proteomics of plant cell membranes, *Journal Experimental Botany* 58: 103–12.

Krishnan, P., Kruger, N.J. & Ratcliffe, R.G. (2005). Metabolite fingerprinting and profiling in plants using NMR. *Journal of Experimental Botany* 56: 255-265.

Li, S.H., Schneider, B. & Gershenzon J. (2007). Microchemical analysis of laser-microdissected stone cells of Norway spruce by cryogenic nuclear magnetic resonance spectroscopy, *Planta* 225: 771-779.

Lippert, D., Chowrira, S., Ralph, S.G., Zhuang, J., Aeschliman, D., Ritland, C., Ritland, K., Bohlmann, J. (2007). Conifer defense against insects: proteome analysis of Sitka spruce (*Picea sitchensis*) bark induced by mechanical wounding or feeding by white pine weevils (*Pissodes strobi*), *Proteomics* 7: 248-70.

Lisec, J., Schauer, N., Kopka, J., Willmitzer, L. & Fernie, A.R. (2006). Gas chromatography mass spectrometry-based metabolite profiling in plants, *Nature Protocols* 1: 387-396.

Lliso, I., Tadeo, F.R., Phinney, B.S., Wilkerson, C.G. & Talón, M. (2007). Protein changes in the albedo of citrus fruits on postharvesting storage, *Journal of Agriculture Food Chemistry* 55: 9047-53.

Lu, P., Vogel, C., Wang, R., Yao, X. & Marcotte, E.M. (2007). Absolute protein expression profiling estimates the relative contributions of transcriptional and translational regulation, *Nature Biotechnology* 25: 117-24.

Martin, C., Bhatt, K. & Baumann, K. (2001). Shaping in plant cells, *Current Opinion in Plant Biology* 4: 540-9.

Matsye, P.D., Kumar, R., Hosseini, P., Jones, C.M., Alkharouf, N., Matthews, B.F.& Klink VP. Mapping cell fate decisions that occur during soybean defense responses. (Submitted)

Matas, A.J., Augusti, J., Tadeo, F.R., Talón, M. & Rose, J.K. (2010). Tissue-specific transcriptome profiling of the citrus fruit epidermis and subepidermis using laser capture microdissection, *Journal of Experimental Botany* 61: 3321-3330.

Mehta, A., Brasileiro, A.C., Souza, D.S., Romano, E., Campos, M.A., Grossi-de-Sá, M.F., Silva, M.S., Franco, O.L., Fragoso, R.R., Bevitori R. & Rocha, T.L. (2008). Plant-pathogen interactions: what is proteomics telling us?, *FEBS Journal* 275: 3731-3746.

Meijers, R., Puettmann-Holgado, R. Skiniotis, G., Liu, J.H., Walz, T., Wang, J.H. & Schmucker, D. (2007). Structural basis of Dscam isoform specificity, *Nature* 449: 487-491.

Michnick, S.W. (2004). Proteomics in living cells, *Drug Discovery Today* 9: 262-267.

Miernyk, J.A., Pret'ova, A., Olmedilla, A., Klubicova, K., Obert B., Hajduch M. et al. (2011). Using proteomics to study sexual reproduction in angioperms, *Sex Plant Reproduction* 24: 9-22.

Moco, S., Bino, R.J., Vorst, O., Verhoeven, H.A., de Groot, J., van Beek, T.A., Vervoort, J. & De Vos, R.C.H. (2006). A liquid chromatography-mass spectrometry-based metabolome database for tomato, *Plant Physiology* 141: 1205-1218.

Moco, S., Capanoglu, E., Tikunov, Y., Bino, R.J., Boyacioglu, D., Hall, R.D., Vervoort, J., De Vos, R.C.H. (2007). Tissue specialization at the metabolite level is perceived during the development of tomato fruit, *Journal Experimental Botany* 58: 4131-4146.

Moco, S., Forshed, J., De Vos, R.C.H., Bino, R.J. & Vervoort, J. (2008). Intra- and inter-metabolite correlation spectroscopy of tomato metabolomics data obtained by liquid chromatography-mass spectrometry and nuclear magnetic resonance, *Metabolomics* 4: 202-215.

Moco, S., Schneider, B. & Vervoort, J. (2009). Plant micrometabolomics: the analysis of endogenous metabolites present in a plant cell or tissue, *Journal of Proteome Research* 8: 1694-1703.

Moulédous, L., Hunt, S., Harcourt, R., Harry, J., Williams, K.L. & Gutstein, H.B. (2002). Lack of compatibility of histological staining methods with proteomic analysis of laser-capture microdissected brain samples, *Journal of Biomolecular Techniques* 13: 258-264.

Moulédous, L., Hunt, S., Harcourt, R., Harry, J., Williams, K.L. & Gutstein, H.B. (2003). Navigated laser capture microdissection as an alternative to direct histological staining for proteomic analysis of brain samples, *Proteomics* 3: 610-615.

Nakazono, M., Qiu, F., Borsuk, L.A. & Schable, P.S. (2003). Laser capture microdissection, a tool for the global analysis of gene expression in specific plant cell types: identification of genes expressed differentially in epidermal cells or vascular tissue of maize, *The Plant Cell* 15: 583-596.

Nelson, C.J., Hegeman, A.D., Harms, A.C. & Sussman, M.R. (2006a). A quantitative analysis of *Arabidopsis* plasma membrane using trypsin-catalyzed 18 O labeling. *Molecular Cell Proteomics* 5: 1382–95.

Nelson, T., Tausta, S.L., Gandotra, N. & Liu, T. (2006b). Laser microdissection of plant tissue: What you see is what you get, *Annual Review of Plant Biology* 57: 181–201.

Nelson, T., Gandotra, N. & Tausta, S.L. (2008). Plant cell types: reporting and sampling with new technologies, *Current Opinion in Plant Biology* 11: 567–573.

Nesatyy, V.J. & Suter, M.J. (2008). Analysis of environmental stress response on the proteome level, *Mass Spectrometry Reviews* 27: 556-74.

Nie, L., Wu, G., Culley, D.E., Scholten, J.C. & Zhang W. (2007). Integrative analysis of transcriptomic and proteomic data: challenges,solutions and applications, *Critical Reviews in Biotechnology* 27: 63–75.

Niittylä, T., Fuglsang, A.T., Palmgren, M.G., Frommer, W.B. & Schulze, W.X. (2007). Temporal analysis of sucrose-induced phosphorylation changes in plasma membrane proteins of *Arabidopsis*, *Molecular Cell Proteomics* 6: 1711–26.

Nowaczyk, M.M., Hebeler, R., Schlodder, E., Meyer, H.E., Warscheid, B. & Rögner, M. (2006). Psb27, a cyanobacterial lipoprotein, is involved in the repair cycle of photosystem II, *The Plant Cell* 18: 3121–3177.

Oda, Y., Huang, K., Cross, F.R., Cowburn, D. & Chait, B.T. (1999). Accurate quantitation of protein expression and site-specific phosphorylation. *Proceedings of the National Academy of Science USA* 96: 6591–96.

O'Farrell, P.H. (1975). High resolution two-dimensional electrophoresis of proteins, *The Journal of Biological Chemistry* 250: 4007-4021.

O'Farrell, P.Z., Goodman, H.M. & O'Farrell P.H. (1977). High resolution two dimensional electrophoresis of basic as well as acidic proteins, *Cell* 12: 1133-1142.

Opitz, S. & Schneider, B. (2002). Organ-specific analysis of phenylphenalenone-related compounds in *Xiphidium caeruleum*, *Phytochemistry* 61: 819–825.

Ohtsu, K., Takahashi, H., Schnable, P.S. & Nakazono, M. (2007). Cell type-specific gene expression profiling in plants by using a combination of laser microdissection and high-throughput technologies, *Plant Cell Physiology* 48: 3-7.

Palmblad, M., Mills, D.J. & Bindschedler, L.V. (2008). Heat-shock response in *Arabidopsis thaliana* explored by multiplexed quantitative proteomics using differential metabolic labeling, *Journal of Proteome Research* 7: 780–85.

Palusa, S.G., Ali. G.S. & Reddy A.S.N. (2007). Alternative splicing of pre-mRNAs of Arabidopsis serine/arginine-rich proteins: regulation by hormones and stresses, *The Plant Journal* 49: 1091–107.

Pan, H. & Lundgren, L.N. (1995). Phenolic extractives from root bark of *Picea abies*, *Phytochemistry* 39: 1423-8.

Pan, Q., Shai, O., Lee, L.J., Frey, B.J. & Blencowe, B.J. (2008). Deep surveying of alternative splicing complexity in the human transcriptome by high-throughput sequencing, *Nature* Genetic 40: 1413-5.

Pedreschi, R.,Vanstreels, E., Carpentier, S., Hertog, M., Lammertyn, J., Robben J., Noben, J.P., Swennen, R., Vanderleyden, J. & Nicolai, B. (2007). Proteomic analysis of core breakdown disorder in conference pears (*Pyrus communis* L.), *Proteomics* 7: 2083-99.

Peltier, J.B., Friso, G., Kalume, D.E., Roepstorff, P., Nilsson, F., Adamskaa, I. & van Wijk K.J. (2000). Proteomics of the chloroplast: systematic identification and targeting analysis of lumenal and peripheral thylakoid proteins, *The Pant Cell* 12: 319–342.

Persidis, A. (1998). Proteomics, *Nature Biotechnology* 4: 393-394.

Pevzner, P.A., Mulyukov, Z., Dancik, V. & Tang, C.L. (2001). Efficiency of database search for identification of mutated and modified proteins via mass spectrometry, *Genome Research* 11: 290–99.

Purea, A., Neuberger, T. & Webb, A.G. (2004). Simultaneous NMR micro-imaging of multiple single-cell samples, *Concepts Magn Reson Part B* 22B: 7–14.

Rabilloud, T. (1996). Solubilization of proteins for electrophoretic analyses, *Electrophoresis* 17: 813-829.

Ramsay, K., Jones, M.G.K. & Wang, Z. (2006). Laser capture microdissection: A novel approach to microanalysis of plant–microbe interactions, *Molecular Plant Pathology* 7: 429–435.

Reiland, S., Messerli, G., Baerenfaller, K., Gerrits, B., Endler, A., Grossmann, J., Gruissem W. & Baginsky, S. (2009). Large-scale *Arabidopsis* phosphoproteome profiling reveals novel chloroplast kinase substrates and phosphorylation networks, *Plant Physiology* 150: 889–903.

Rekhter, M.D. & Chen, J. (2001). Molecular analysis of complex tissues is facilitated by laser capture microdissection: Critical role of upstream tissue processing, *Cell Biochemistry and Biophysics* 35: 103–113.

Roessner, U., Willmitzer, L. & Fernie, A.R. (2001). High-resolution metabolic phenotyping of genetically and environmentally diverse potato tuber systems. Identification of phenocopies, *Plant Physiology* 127: 749–764.

Rossignol, M., Peltier, J.B., Mock, H.P., Matros, A., Maldonado, A.M. & Jorrin J.V. (2006). Plant proteome analysis: a 2004–2006 update, *Proteomics* 6: 5529-48.

Rubakhin, S.S., Greenough, W.T. & Sweedler, J.V. (2003). Spatial profiling with MALDI MS: distribution of neuropeptides within single neurons, *Analytical Chemistry* 75: 5374-5380.

Salekdeh, G.H. & Komatsu, S. (2007). Crop proteomics: aim at sustainable agriculture of tomorrow, *Proteomics* 7: 2976–96.

Schad, M,. Lipton, M.S., Giavalisco, P., Smith, R.D. & Kehr, J. (2005a). Evaluation of two-dimensional electrophoresis and liquid chromatography–tandem mass spectrometry for tissue-specific protein profiling of laser-microdissected plant samples, *Electrophoresis* 26: 2729–2738.

Schad, M., Mungur, R., Fiehn, O. & Kehr, J. (2005b). Metabolic profiling of laser microdissected vascular bundles of *Arabidopsis thaliana*, *Plant Methods* 1:2.

Schmucker, D., Clemens, J.C., Shu, H., Worby, C.A., Xiao, J., Muda, M., Dixon, J.E. & Zipursky, S.L. (2000). *Drosophila* Dscam is an axon guidance receptor exhibiting extraordinary molecular diversity, *Cell* 101: 671-684.

Schneider, B. & Hölscher, D. (2007). Laser microdissection and cryogenic nuclear magnetic resonance spectroscopy: an alliance for cell type-specific metabolite profiling, *Planta* 225: 763-770.

Schulze, W.X. & Usadel, B. (2010). Quantitation in mass-spectrometry-based proteomics, *Annual Review in Plant Biology* 61: 491-516.

Sheoran, I.S., Ross, A.R., Olson, D.J. & Sawhney, V.K. (2007). Proteomic analysis of tomato (*Lycopersicon esculentum*) pollen, *Journal of Experimental Botany* 58: 3525-35.

Shibutani, M., Uneyama, C., Miyazaki, K., Toyoda, K. & Hirose, M. (2000). Methacarn Fixation: A Novel Tool for Analysis of Gene Expressions in Paraffin-Embedded Tissue Specimens. *Lab Invest.* 80: 199-208.

Simon, S.A., Zhai, J., Nandety, R.S., McCormick, K.P., Zeng, J., Mejia, D. & Meyers, B.C. (2009). Short-Read Sequencing Technologies for Transcriptional Analyses, *Annual Review of Plant Biology* 60: 305-33.

Simpson, C.G., Lewandowska, D., Fuller, J., Maronova, M., Kalyna, M., Davidson, D., et al. (2008) Alternative splicing in plants, *Biochem Soc Trans* 36: 508-10.

Sitek, B., Luttges, J., Marcus, K., Kloppel, G., Schmiegel, W., Meyer, H. E., Hahn, S. A. & Stuhler, K. (2005). Application of fluorescence difference gel electrophoresis saturation labelling for the analysis of microdissected precursor lesions of pancreatic ductal adenocarcinoma, *Proteomics* 5: 2665-2679.

Song, X., Ni, Z., Yao, Y., Xie, C., Li, Z., Wu, H, Zhang, Y. & Sun, Q. (2007). Wheat (*Triticum aestivum* L.) root proteome and differentially expressed root proteins between hybrid and parents, *Proteomics* 27: 3538-57.

Sturm, S., Seger, C., Godejohann, M., Spraul, M. & Stuppner, H. (2007). Conventional sample enrichment strategies combined with high- performance liquid chromatography-solid phase extraction-nuclear magnetic resonance analysis allows analyte identification from a single minuscule *Corydalis solida* plant tuber, *Journal of Chromatography* 1163: 138-144.

Takáč, T., Pechan, T. & Samaj, J. (2011). Differential proteomics of plant development, *Journal of Proteomics* 74: 577-88.

Takahashi, H., Kamakura, H., Sato, Y., Shiono, K., Abiko, T., Tsutsumi, N., Nagamura, Y., Nishizawa, N.K., Nakazono, M. (2010). A method for obtaining high quality RNA from paraffin sections of plant tissues by laser microdissection, *Journal of Plant Research* 123: 807-813.

Tanabe, N., Yoshimura, K., Kimura, A., Yabuta, Y. & Shigeoka, S. (2007). Differential expression of alternatively spliced mRNAs of *Arabidopsis* SR protein homologs, atSR30 and atSR45a, in response to environmental stress, *Plant Cell Physiology* 48: 1036-1049.

Tang, W., Coughlan, S., Crane, E., Beatty, M. & Duvick, J. (2006). The application of laser microdissection to in planta gene expression profiling of the maize anthracnose stalk rot fungus *Colletotrichum graminicola*, *Molecular Plant–Microbe Interaction* 19: 1240-1250.

Tanner, S., Shu, H., Frank, A., Wang, L.C., Zandi, E., Mumby, M., Pevzner, P.A. & Bafna, V. (2005). InsPecT: identification of posttranslationally modified peptides from tandem mass spectra, *Analytical Chemistry* 77: 4626-39.

Thiel, J., Weier, D., Sreenivasulu, N., Strickert, M., Weichert, N., Melzer, M., Czauderna, T., Wobus, U., Weber, H., Weschke, W. (2008) Different hormonal regulation of cellular differentiation and function in nucellar projection and endosperm transfer cells—a microdissection-based transcriptome study of young barley grains, *Plant Physiology* 148: 1436–1452.

Thiel, J., Müller, M., Weschke, W. & Weber, H. (2009). Amino acid metabolism at the maternal–Wlial boundary of young barley seeds: a microdissection-based study, *Planta* 230: 205–213.

Valledor, L., Castillejo, M.A., Lenz, C., Rodríguez, R., Cañal, M.J. & Jorrín, J. (2008). Proteomic analysis of Pinus radiata needles: 2-DE map and protein identification by LC/MS/MS and substitution-tolerant database searching, *Journal of Proteome Research* 7: 2616–31.

van der Weerd, L., Claessens, M.M.A.E., Efde, C. & Van As, H. (2002). Nuclear magnetic resonance imaging of membrane permeability changes in plants during osmotic stress, *Plant Cell and Environment* 25: 1539-1549.

Vyetrogon, K., Tebbji, F., Olson, D.J., Ross, A.R. & Matton, D.P. (2007). A comparative proteome and phosphoproteome analysis of differentially regulated proteins during fertilization in the self-incompatible species *Solanum chacoense* Bitt., *Proteomics* 7: 232–47.

Wang, Y.L., Tang, H.R., Nicholson, J.K., Hylands, P.J., Sampson, J., Whitcombe, I., Stewart, C.G., Caiger, S., Oru, I., Holmes, E. (2004). Metabolomic strategy for the classification and quality control of phytomedicine: a case study of chamomile flower (*Matricaria recutita* L.), *Planta Medica* 70: 250-5.

Wang, D., Eyles, A., Mandich, D. & Bonello, P. (2006). Systemic aspects of host–pathogen interactions in Austrian pine (*Pinus nigra*): a Proteomics approach. *Physiological and Molecular Plant Pathology* 68: 149–57.

Ward, J.L., Harris, C., Lewis, J. & Beale, M.H. (2003). Assessment of H-1 NMR spectroscopy and multivariate analysis as a technique for metabolite fingerprinting of *Arabidopsis thaliana*, *Phytochemistry* 62: 949–957.

Wasinger, V.C., Cordwell, S.J., Cerpa-Poljak, A., Yan, J.X., Gooley, A.A., Wilkins, M.R., Duncan, M.W., Harris, R., Williams, K.L. & Humphery-Smith, I. (1995). Progress with gene-product mapping of the Mollicutes: *Mycoplasma genitalium*, *Electrophoresis* 16: 1090-1094.

Wenzler, M., Hölscher, D., Oerther, T. & Schneider, B. (2008). Nectar formation and floral nectary anatomy of *Anigozanthos flavidus*: a combined magnetic resonance imaging and spectroscopy study, *Journal of Experimental Botany* 59: 3425-3434.

Witze, E.S., Old, W.M., Resing, K.A. & Ahn N.G. (2007). Mapping protein post-translational modifications with mass spectrometry, *Nature Methods* 4: 798–806.

Xiao, Y.L., Smith, S.R., Ishmael, N., Redman, J.C., Kumar, N., Monaghan, E.L., Ayele, M., Haas, B.J., Wu, H.C. & Town C.D. (2005). Analysis of the cDNAs of hypothetical genes on *Arabidopsis* chromosome 2 reveals numerous transcript variants, *Plant Physiology* 139: 1323-37.

Yoshimura, K., Masuda, A., Kuwano, M., Yokota, A. & Akashi, K. (2008). Programmed proteome response for drought avoidance/tolerance in the root of a C3 xerophyte (wild watermelon) under water deficits, *Plant and Cell Physiology* 49: 226–41.

Yu, Y.Y., Lashbrook, C. & Hannapel, D. (2007). Tissue integrity and RNA quality of laser microdissected phloem of potato, *Planta* 226: 797–803.

# 5

# Simple and Rapid Proteomic Analysis by Protease-Immobilized Microreactors

Hiroshi Yamaguchi[1,2], Masaya Miyazaki[1,3] and Hideaki Maeda[1,3]
*¹Measurement Solution Research Center,*
*National Institute of Advanced Industrial*
*Science and Technology*
*²Liberal Arts Education Center,*
*Aso campus, Tokai University*
*³Interdisciplinary Graduate School of*
*Engineering Science, Kyusyu University*
*Japan*

## 1. Introduction

Proteomics is the large-scale study of proteins, particularly their stuructures and functions. One of the most important point is to develop efficient and rapid approachehes to identify the target proteins. Peptide mass mapping and tandem mass spectrometry (MS/MS)-based peptide sequencing are key methods in current protein identification for proteomic studies. Proteins are usually digested into peptides that are subsequently analyzed by MS. Therefore, proteolysis by sequence-specific proteases is the key step for positive sequencing in proteomic analysis integrated with MS (Aebersold & Mann, 2003). The conventional method of in-solution digestion by proteases is a time-consuming procedure (overnight at 37 °C). The substrate/protease ratio must be kept high (generally > 50) in order to prevent excessive sample contamination by the protease and its auto-digested products. But this leads to a relatively slow digestion. In addition, obtaining reliable peptide maps and meaningful sequence data by MS analysis requires not only the separation of the digested peptides but also strictly defined proteolysis conditions (Domon & Aebersold, 2006; Witze et al., 2007). In addition, the ionization efficiency of the digested fragments including a modified peptide such as a phosphopeptide is dependent on peptide-size or peptide-sequence, which directly correlates with sequence coverage of the target protein by MS analysis. Furthermore, peptide recovery from in-solution digestion is highly dependent on the structural properties of the target proteins because proteins with rigid structures, *e.g.* by disulfide bonds tend to be resistant to complete digestion. In fact, the typical preparation of a sample for proteolysis includes denaturation, reduction of disulfide bonds, and alkylation procedures lead to a decrease the conformational stability. It is obvious that insufficient sequence coverage could compromise the accuracy of proteome characterization. Therefore, it is important to develop novel digestion methods to achieve a highly efficient proteolysis for MS-based peptide mapping (Park & Russell, 2000; Slysz et al., 2006).

A microreactor is a suitable reaction system for handling small-volume samples in a microchannel to perform chemical or enzymatic reactions. Enzyme-immobilized microreactors have been widely utilized in chemical and biotechnological fields (Liu et al., 2008; Ma et al., 2009; Miyazaki et al., 2008; Asanomi et al., 2011). The protease-immobilized microreactor provides several advantages for proteolysis (Ma et al., 2009); *e.g.* low degree of auto-digestion even at high protease concentrations and a large surface and interface area that leads to rapid proteolysis. Furthermore, the immobilized proteases on the microchannel walls can be easily isolated and removed from the digested fragments prior to MS, which means elimination of the requirement to stop the reaction by chemical or thermal denaturation after digestion. These features can contribute to higher sequence coverage compared to the approach based on in-solution digestion. High sequence coverage is important to enhance the probability of identification of the protein and increase the likelihood of detection of structural variants generated by processes such as post-translational modifications.

Several methods for protease immobilization have been reported, wherein the protease, usually trypsin, has been immobilized in microchips by sol-gel encapsulation (Sakai-Kato et al., 2003; Wu et al., 2004), covalently bounded (Lee et al., 2008; Fan & Chen, 2007) and physically adsorbed onto different supports (Liu et al., 2006). In addition, trypsin-immobilized magnetic particles have been developed to carry out proteolysis with a short digestion time (Li et al., 2007a; Chen & Chen, 2007). However, preparations of these protease-immobilized microreactors require multi-step procedures consuming considerable amounts of time and effort. Therefore, a facile preparation method of the enzyme-immobilized microreactor is desirable for the routine proteolysis step in proteomic analysis. In addition, reusability is also an important feature required for laboratory use.

We developed the procedure for immobilizing enzymes on the internal surface of the poly-tetrafluoroethylene (PTFE) microtube by forming an enzyme polymeric membrane through a cross-linking reaction in a laminar flow between lysine (Lys) residues on the protein surfaces (Honda et al., 2005) or between the mixture of proteins with isoelectric point p$I$ < 7.0 and poly-Lys (Honda et al., 2006). A typical sample preparation for proteolysis involves multi-steps (denaturation, reduction and alkylation) that are expected to produce enhancement of digestion efficiency. Because enzyme-immobilized microreactors prepared by our cross-linking method have excellent reaction performance and stability against high temperature and high concentration of denaturant (Honda et al., 2005), the microreactors are expected to achieve efficient digestion during the denaturation process. This idea inspired us to apply protease-immobilized microreactor for rapid and accurate proteolysis in proteomic analysis.

This chapter addresses the use of protease-immobilized microreactors with MS for proteomic applications. Preparation of the microreactors and examples of a simple and rapid analysis of protein sequence and protein post-translational modification are presented.

## 2. Preparation of protease-immobilized microreactors

A typical procedure for cross-linking enzyme involves activation of the primary amine groups of enzyme with cross-linker to create aldehyde groups that can react readily with other primary amine groups of enzymes (Honda et al; 2005; Ma et al., 2009; Miyazaki & Maeda, 2006). Cross-linking yields depend on the number of the Lys residues of enzyme. Therefore, the acidic or neutral enzyme (p$I$ < 7.0) cannot be cross-linked efficiently merely by the use of cross-linker such as glutaraldehyde (GA). Although high concentration of cross-linker can increase the cross-linking yield, it often causes a change of its conformation, engendering a reduction of its catalytic activity (Wang et al., 2009). To overcome this difficulty, poly-Lys was used as the amine donor in this study. It is expected that the large number of primary amine groups of poly-Lys can improve the cross-linking yield with lower concentration of cross-linker than those in reported procedures (typically 5–10% of GA, v/v) (Ma et al., 2008; Fan & Chen, 2007).

## 2.1 Protease
We prepared two protease-immobilized microreactors for proteomic analysis; trypsin- (TY) and chymotrypsin- (CT) immobilized microreactors. TY hydrolyses peptide bonds after Arg and Lys residues. Because these basic residues are usually located on the surface of protein, especially in soluble proteins, the digested peptides generally fit the range (< 2 kDa) required for analysis by MS. However, if Pro residue is located at the C-terminal side of Arg or Lys, hydrolysis will not occur. Moreover, it is possibility that the conformational stability of the protein *e.g.* by disulfide bond has a resistance for proteolysis. These possibilities will cause the digested peptides to become too large to be detected by MS. Therefore, aside from TY, other endoproteases can be used for MS-based analysis to cover the whole sequence of the target protein. CT hydrolyses peptide bonds after aromatic residues (Phe, Trp and Tyr) and after Leu and Met in a less specific way and is often used for proteolysis of hydrophobic proteins such as membrane proteins (Fischer & Poetsch, 2006; Temporini et al., 2009).

## 2.2 Preparation techniques for the protease-immobilized microreactors
A microfluidics-based enzyme-polymerization technique (Honda et al., 2005; Honda et al., 2006) was used for the preparation of protease-immobilized microreactors. The procedure for immobilizing protease on the internal surface of the PTFE microtube by forming an enzyme polymeric membrane through a cross-linking reaction between Lys residues on the protein surface are presented in Fig. 1. The solutions were introduced to the PTFE microtube from gas tight syringes by syringe pumps. The combination of GA (0.25%, v/v) and paraformaldehyde (PA) (4%, v/v) provided better cross-linking yields between the proteases maintaining their activities (Honda et al., 2005; Honda et al., 2006). Because the p$I$ value of CT (8.6) is close to the pH value of the reaction buffer (8.0), this probably leads to low polymerization yield. Thus, poly-Lys supporting polymerization procedure was used for the preparation of CT-immobilized microreactor. The molecular weight of poly-Lys was 62 kDa. Resulting Schiff base was reduced by NaCNBH$_3$. On the other hand, for TY-immobilized microreactor, poly-Lys was omitted because the p$I$ value of trypsin was 10.5 and poly-Lys was a substrate for trypsin. To avoid autolysis of protease in bulk solution during the cross-linking reaction, the preparation of microreactor was conducted at 4 °C (Yamaguchi et al., 2009).

Protease immobilization on PTFE tube was analyzed by the Bradford method in order to measure the total polymerized enzyme. For example, 50 µg CT was formed by polymerizing on a 1 cm long PTFE tube. Other enzymes were also immobilized on PTFE tubes with similar concentration. Because these concentrations of the immobilized proteases are higher than those used in the experimental conditions of in-solution digestion, it can be suggested that our microreactors can perform rapid digestion compared with in-solution digestion.

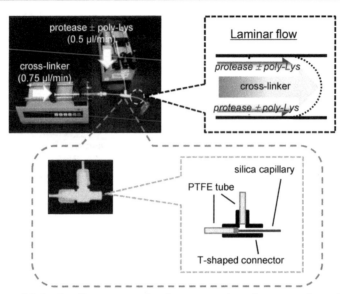

Fig. 1. The assembled microflow system for the preparation of protease-immobilized microreactor (Honda et al., 2006; Yamaguchi et al., 2009). The cross-linker solution was supplied to the substrate PTFE microtube through a silica capillary, corresponding to a central stream in the concentric laminar flow. A solution of proteases or a protease/poly-Lys mixture was supplied from another PTFE microtube connected to the T-shaped connector. Both solutions were introduced by syringe pumps.

In addition to these protease-immobilized microreactors, we also prepared alkaline phosphatase- (AP) immobilized microreactor for analysis of protein phosphorylation. Because the p$I$ value of AP is 5.9, the poly-Lys supported immobilization procedure was used for the AP-immobilized microreactor. When high molecular weight of poly-Lys (62 kDa) which was intended for CT-microreactor was used for the AP-microreactor, an aggregation of protein was readily observed, suggesting that high positively charged poly-Lys (62 kDa) was quickly interacted with the acidic AP protein by electrostatic interaction. Because our enzyme-polymeric membrane is formed on the inner wall of the microchannel (500 µm inner diameter) through cross-linking polymerization in a laminar flow (Fig. 1), the quickly aggregated enzyme and poly-Lys can be stucked on the microchannel during cross-linking polymerization. Therefore, it is suggested that the large poly-Lys (62 kDa) molecular is not appropriate for preparation of AP-microreactor. To overcome this problem, a low molecular weight poly-Lys (4 kDa) was used for the polymerization of AP. As expected, with the use of poly-Lys (4 kDa), quick aggregation was suppressed and AP-microreactor was successfully prepared.

## 3. Characterization of the protease-immobilized microreactors

### 3.1 Kinetic characterization

In our digestion procedure, the substrate solution was pumped through the microreactor using a syringe pump (Fig. 2A). A reaction time is correlated with a flow rate of the substrate. In the present microreactors, the hydrolysis reaction at a flow rate of 5.0 µl/min

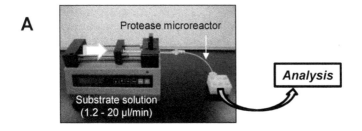

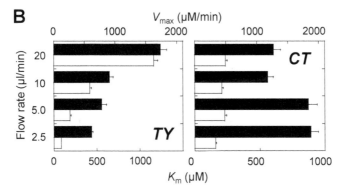

Fig. 2. (A) Schematic representation of the hydrolytic reaction by the protease-immobilized microreactor. Reaction temperature was kept in an incubator. (B) Kinetic parameters of hydrolysis activity of the protease-immobilized microreactors at different flow rates of the substrates. Substrates: BAPA for TY; GPNA for CT. Open bars, $K_m$ (µM); closed bars, $V_{max}$ (µM/min). The graph shows the mean ± standard error for at least three experiments. All assays were performed in 50 mM Tris–HCl (pH 8.0) at 30 °C.

yields a reaction time of 5.2 min (PTFE microtube volume of 26 µl). The individual hydrolysis activity of the microreactors was evaluated using synthetic small compounds; benzoyl-L-arginine *p*-nitroanilide (BAPA) for TY-microreactor and *N*-glutaryl-L-phenylalanine *p*-nitroanilide (GPNA) for CT-microreactor. Digestions by both microreactors showed a similar order of $K_m$ values (hundred micromolar) to the reported $K_m$ values for TY in a microtube (Yamashita et al., 2009) and CT-microreactor (Honda et al., 2005), suggesting that both immobilized proteases maintain their own hydrolysis activity after polymerization on PTFE surface.

It is known that the flow rate of substrate can affect the efficiency of the immobilized enzyme activity (Fan & Chen, 2007; Honda et al., 2005; Ma et al., 2008; Nel et al., 2008; Wu et al., 2004; Dulay et al., 2005). Therefore, we studied the effect of different delivery speeds of substrates on hydrolysis activities (Yamaguchi et al., 2009). As shown in Fig. 2B, the estimated $K_m$ values for TY-microreactor decreased with increase in flow rate, while no significant change in $K_m$ values was observed for CT-microreactor at different flow rate, indicating that the diffusion limitation of the substrate in the immobilized TY is more influenced than that in the immobilized CT. Because both proteases have similar conformational structures and catalytic-sites as well-known, the difference between TY-microreactor and CT-microreactor could be attributed to the difference in polymerization procedure. In contrast to $K_m$ values, $V_{max}$ values for both microreactors increased with increase in flow rate. Therefore, both microreactors showed higher $V_{max}/K_m$ values at lower flow rate (longer reaction time) but lower $V_{max}/K_m$ values at higher flow rate (shorter reaction time), suggesting that the hydrolysis activity was more efficient at lower flow rate. This indicates that efficient digestion of substrate by immobilized protease was achieved using an alternative substrate mobilization approach that involved incubation of substrate with immobilized enzyme. In contrast, the calculated $K_m$ value for free TY was 806 μM, a $K_m$ value that was 4.3-fold higher than TY-microreactor with the same reaction time. Free CT also showed much higher $K_m$ value (over 1 mM) compared with that of CT-microreactor. It is known that $K_m$ value represents the binding affinity between enzyme and substrate, and lower $K_m$ value means higher affinity therefore, suggesting that the enhanced mass-transfer in the microchannel induced hydrolysis reaction (Honda et al., 2005; Honda et al., 2007).

## 3.2 Operational stabilities of the protease-immobilized microreactors

The operational stability and reusability of the protease-immobilized microreactor are important for its application in the proteomics analysis. Therefore, the stability of immobilized proteases against temperature in the hydrolysis reaction were tested (Yamaguchi et al., 2009). The relative hydrolysis activity against BAPA or GPNA of the protease-immobilized microreactors and free proteases at different temperatures were measured. The results showed that both immobilized proteases were more stable at high temperature than free proteases. At 50 °C, free TY and free CT showed 15 and 52% of hydrolysis activities respectively, while the immobilized proteases kept at 30 °C retained their activities. It is suggested that due to multipoint interactions between the TY molecules or between CT and poly-Lys, the immobilized proteases increased the thermal stabilities of the enzymes. Similar thermal stability of immobilized enzyme was previously reported (Kim et al., 2009; Sheldon 2007).

Next, the stability of the protease-microreactors against high concentration of denaturant was tested. When the substrate solutions in 3 M guanidine hydrochloride (Gdn-HCl) were delivered to both protease-immobilized microreactors, we observed almost the same hydrolysis activities compared with those in buffer containing no denaturant. In contrast, free proteases in 3 M Gdn-HCl were denatured consequently and did not show any hydrolysis activities. Similar stability of CT-microreactor against 4 M urea and 50% dimethyl sulfoxide were previously observed (Honda et al., 2005). Although both immobilized proteases did not show any hydrolysis activities in 5 M Gdn-HCl, over 90% recovered activities were observed after washing the microreactors with buffer containing no Gdn-HCl. Moreover, the reusability of the microreactors was also investigated by storing them at pH 8.0 at 4 °C. After over 60 days (after over 20 times reuse), both microreactors retained over 90% of their hydrolysis activities against synthetic substrates while free proteases had very little to no activities. More recently, the immobilized subtilisin by poly-Lys supported cross-linking was also more stable than free protease at high temperature, in the presence of a chemical denaturant or in an organic solvent and was recycled without appreciable loss of activity (Yamaguchi et al., 2011). These results indicate that the stability of proteases was improved after formation of enzyme-polymerization. This enhancement in the efficiency of activity of the immobilized protease compared to free protease can be ascribed to minimization or elimination of the auto-digestion of proteases (Dulay et al., 2005; Shui et al., 2006) and possible stabilization of the structure of proteases by cross-linking, thus resulting in higher accessibility of the substrate to the catalytic-site of the enzyme (Honda et al., 2005; Honda et al., 2006; Dulay et al., 2005; Shui et al., 2006).

The operational stability of the protease-immobilized microreactors was also tested based on the digestions of cytochrome $c$ (Cyt-C) for TY-microreactor and β-casein for CT-microreactor (Yamaguchi et al., 2010a). Between each digestion, both microreactors were washed with buffer solution and stored at 4 °C for over 60 days. The obtained 10 MS spectra (not shown) were identical with the similar sequence coverage of 93% (Cyt-C by TY-microreactor) and 57% (β-casein by CT-microreactor). In contrast, free proteases almost completely lost their activities at 25 °C within a couple of days, as reported previously (Sakai-Kato et al., 2002). These results indicate that the stability of proteases was increased by the prevention of auto-digestion after an enzyme-immobilization.

### 3.3 Proteolysis by the protease-immobilized microreactors

Our previous proteolysis procedure using microreactors was carried out in 50 mM Tris–HCl (pH 8.0) solution (Yamaguchi et al., 2009). In this buffer system, an additional purification step using reversed-phase micropipette tips prior to MS measurement could remove excessive amounts of buffer salt, but could lead to sample losses especially of hydrophobic peptides due to their inherent affinity to reversed-phase surfaces, leading to lower sequence coverage. To avoid this, proteolysis was carried out in a 10 mM ammonium acetate buffer (pH 8.5) that easily evaporated during ESI-TOF MS measurement without the need for any desalting procedure. The digested peptides were collected in a test tube and then directly analyzed by ESI-TOF MS using Mariner mass spectrometer (Applied Biosystems Inc.) and reverse-phase high performance liquid chromatography (HPLC) (Yamaguchi et al, 2010a).

| Digestion methods | Temperature /°C | Reaction time | Identified amino acids | Sequence coverage / % |
|---|---|---|---|---|
| TY-microreactor | 30 | 10.4 min | 97/104 | 93 |
| CT-microreactor | 30 | 10.4 min | 40/104 | 38 |
| TY (in-solution)[a] | 37 | 18 hours | 99/104 | 95 |
| CT (in-solution)[a] | 37 | 18 hours | 68/104 | 65 |

[a]In-solution digestion was carried out in 10 mM ammonium acetate buffer, pH 8.5. Concentrations of substrate and free proteases were 100 and 2 μg/ml, respectively.

Table 1. Summary of ESI-TOF MS results of the digests of Cyt-C

To study the effect of flow rate on the proteolysis efficiency, we first investigated the digestion of Cyt-C by TY-microreactor at several flow rates. Under our experimental conditions, with flow rate increase from 1.2 to 15 μl/min, auto-digestion peak of protease was not observed by MS and HPLC analyses. In addition, MALDI-TOF MS analysis using Bruker Autoflex (Bruker Daltonics) indicated that no free protease or cross-linked aggregation came off from the PTFE tubes, demonstrating good mechanical stability. Proteolysis was carried out at 30 °C. The digests by TY-microreactor were analyzed by ESI-TOF MS spectra and reverse-phase HPLC profiles. With a flow range of 2.5–15 μl/min, the intact Cyt-C was observed while below 1.2 μl/min over 90% of Cyt-C was digested by TY-microreactor. The results indicate that digestion at a lower flow rate (longer reaction time) is more efficient as shown in Fig. 2B (against small compound). Although intact Cyt-C remained in the samples above 2.5 μl/min flow rate, the matched peptides covered 93%

(97/104 amino acids) of the Cyt-C sequence (Table 1). This value was the same as that of the digested sample at 1.2 μl/min. For comparison, in-solution digestions of Cyt-C with an incubation time of 18 hours at 30 and 37 °C were performed. Although the reaction at 37 °C showed complete digestion, only 40% Cyt-C was digested at 30 °C, which is of the same proteolysis condition as that of the TY-microreactor. Thus, these results indicate that digestion of Cyt-C by TY-microreactor was much faster and efficient than in-solution digestion and that there is no need to perform the digestion at higher temperature. In contrast to TY-microreactor, CT-microreactor showed lower sequence coverage (38%) than that by in-solution digestion (65%). Because Cyt-C was not denatured by denaturant in this digestion condition, a possible reason for the lower digestion ability of CT-microreactor could be mass transfer limitation of folded Cyt-C in the cross-linked CT and poly-Lys complex matrix.

The residues of Lys and Arg which are recognized by TY, are hydrophilic residue that usually locate the surface of protein, while aromatic residues for CT are buried inside the protein. It is possible that immobilized TY could easily recognize the residues of Lys and Arg; on the other hand, immobilized CT could not interact with aromatic residues. As described above, the microreactors have stability against high concentration of denaturant and the denatured Cyt-C with its aromatic residue exposed to the outside is expected to be digested by CT-microreactor. To confirm this possibility, Cyt-C digestion by CT-microreactor was carried out in 3 M Gdn-HCl. Circular dichroism (CD) spectrum revealed that Cyt-C (100 μg/mL) in 3 M Gdn-HCl was denatured. Under this condition, free CT did not show any proteolysis activity. In contrast to in-solution digestion, the digests by CT-microreactor showed a different HPLC profile compared with the HPLC profile of intact Cyt-C in 3M Gdn-HCl, indicating that the immobilized CT in 3M Gdn-HCl digested the denatured Cyt-C (Yamaguchi et al., 2009). Furthermore, MS analysis of the digests also showed the efficient digestion by CT-microreactor in 3M Gdn-HCl. These results suggest that the stability of immobilized proteases is superior to that of free proteases.

A typical sample preparation for proteolysis before digestion involves multi-steps including denaturation, reduction of disulfide bond, and alkylation of free thiol group to reduce the conformational stability of protein; steps which are expected to produce enhancement of digestion efficiency (Ma et al., 2008; Li et al., 2007a; Ethier et al., 2006; Lin et al., 2008). However, the multi-step procedure is time-consuming. In contrast, our digestion method by protease-immobilized microreactor can be carried out with high concentration of denaturant and high sequence coverage. Moreover, it can directly use the denatured protein as a substrate. In addition, it does not need any complicated reduction and alkylation steps, therefore, exhibiting superior advantages of our digestion procedure over other reported protease-immobilized microreactor in achieving rapid proteomics analysis.

### 3.4 Proteolysis by the tandem protease-immobilized microreactors

The improved sequence coverage is important to enhance the probability of identification and increase the likelihood of detection of structural variants generated by processes such as alternative splicing and post-translational modifications. The identification of the protein sequence is a first and important step in proteome analysis. Post-translational modification such as phosphorylation is also important information in understanding the role of target protein in the regulation of fundamental cellular processes. Because disregulations of mechanisms of modifications are implicated in various diseases, including cancer (Hunter, 2009), the characterization of protein sequence is useful for biological and clinical researches.

The conventional approach using multi-digestion by different proteases for improved sequence coverage is based on parallel digestions of the same samples and analyzes overlapping peptides. However, this approach takes long time and multi-step procedure. To overcome this difficulty, we prepared the tandem microreactor that was connected by different protease-immobilized microreactors using a Teflon connector. Connection was made easy because the present microreactors were made of PTFE microtube. It is expected that the combination of MS results obtained with the tandem microreactor that carries out multi-digestion may give significantly higher sequence coverage than that obtained with individual digestion by the single microreactor. Based on a similar idea, a reactor which was a bonded mixture of TY and CT to an epoxy monolithic silica column was reported (Temporini et al., 2009). In contrast to the mixture of TY and CT immobilized reactor (Temporini et al., 2009), an interesting feature of our microreactor is the ease in linking each microreactor by using a connector. In our system, we can easily change the order of each microreactor (for example, CT-TY or TY-CT) according to our preference (Yamaguchi et al., 2010a).

Cyt-C (non-phosphoprotein), β-casein (phosphoprotein) and pepsin A (phosphoprotein) were used to test the performance of the tandem microreactors. The enzymatic reaction at a flow rate of 2.5 μl/min yielded a reaction time of 10.4 min (single microreactor) or 20.8 min (tandem microreactor). The digestion efficiency by the microreactors was evaluated by analyzing the sequence coverage and the identified peptide. Some of the digested peptides by the microreactors were the expected peptides that have one or two missed cleavage site. The p*I* value of Cyt-C is 9.6 (horse residue 2-104), suggesting that Arg or Lys residues locate the protein surface with high possibility. As shown above, Cyt-C was digested by TY-microreactor with higher sequence coverage (93% in Table 1) at 30 °C. MALDI-TOF MS analysis also showed high sequence coverage (89%). The value of sequence coverage was higher than the other trypsin-immobilized reactors reported (Li et al., 2007a; Liu et al., 2009) and the same as that performed by in-solution digestion (37 °C for 18 hours), suggesting that the immobilized TY showed rapid and efficient proteolysis.

The multi-digestion of Cyt-C by the tandem microreactor that was connected by using TY-microreactor and CT-microreactor was evaluated. TY-CT tandem microreactor also showed rapid Cyt-C digestion with the similar sequence coverage as that of the single TY-microreactor. As expected, the multi-digested peptides such as VQK, TGPNLHGLF or TGQAPGF were identified (Yamaguchi et al., 2010a). Because some digested peptides (< 300 Da) by tandem microreactor were too small to be identified by TOF-MS analysis, therefore based on comparison with the single TY microreactor the sequence coverage for Cyt-C by the tandem microreactor was found not to improve (88%, 91/104 amino acids). These results indicate that the peptide fragments by TY-microreactor were also digested by CT-microreactor, as expected.

Bovine β-casein (residue 16-224) is a phosphoprotein with well-characterized phosphorylated sites (Han et al., 2009; Li et al., 2007b). MS measurement of an intact protein revealed that β-casein in this study has five phosphorylated sites. By measuring the decrease in the peak area at 220 nm of HPLC profiles of the digests, it could be estimated that over 90% β-casein was digested by CT-microreactor for 10.4 min at 30 °C. ESI-TOF MS analysis revealed that 13 peptides containing 120 out of the 209 possible amino acids of β-casein were obtained, producing the sequence coverage of 57% (Yamaguchi et al., 2010a). This value was higher than that by in-solution digestion (45%, 95/209 amino acids). Moreover, the phosphopeptide containing four phosphoserine (pS) residues (**pSpSpSEEpSITRINKKIEKF**) was detected despite the low ionization efficiency of the phosphopeptide. To confirm this detection of the phosphopeptide by other MS system, MALDI-TOF MS analysis was performed. In addition to the **pSpSpSEEpSITRINKKIEKF** phosphopeptide, the QpSEEQQQTEDEL phosphopeptide was also identified by MALDI-TOF MS analysis, thereby showing that all phosphorylation sites on β-casein in this study were detected from the digests by CT-microreactor. In contrast, HPLC analysis of digested β-casein by TY-microreactor showed a broad profile and was different with that by CT-microreactor. In addition to the p*I* value of β-casein (5.1) that was estimated from the primary structure and which did not take into account the number of phosphorylation sites, the total number of Phe, Trp, Tyr, Leu and Met residues (42) was larger than that of Arg and Lys residues (15). Therefore, it is suggested that the sequence coverage of β-casein by TY-microreactor (14%) or in-solution digestion by free trypsin (21%) is lower than that by CT-microreactor. We next studied the feasibility of enzyme-immobilized tandem microreactor. As expected, the multi-digestion by CT-TY tandem microreactor showed 20 digested peptides (Yamaguchi et al., 2010a). It is noteworthy that GVSK, VKEAMAPK, HKEMPFPK and YPVEPF peptides that were not identified by the single CT-microreactor were identified by the tandem microreactor. The results indicate that an improvement of the sequence coverage in digestion by tandem microreactor in comparison to the single microreactor and in-solution digestion can be expected.

### 3.5 Phosphorylation site analysis by the tandem microreactor

Further analysis of protein phosphorylation was carried out using AP-microreactor. HPLC profile of the digested peptides by CT-AP tandem microreactor revealed the disappearance of two peaks compared to that by the single CT-microreactor. This suggested that the phosphopeptides containing **pSpSpSEEpSITRINKKIEKF** were dephosphorylated by AP. MS analysis also revealed the dephosphorylation of the **pSpSpSEEpSITRINKKIEKF** phosphopeptide. The results indicate that the tandem microreactor which was made by using the protease-microreactor and the phosphatase-microreactor showed the feasibility of the identification of phosphorylation site in phosphoproteins without any enrichment strategies and radioisotope labeling. β-Casein has another well-known phosphorylated site (Ser35). Because the phosphopeptide containing Ser35 (QpSEEQQQTEDEL) was detected by MALDI-TOF MS analysis, it is possible that the one peak that disappeared in HPLC profile may be the phosphopeptide containing pS35.

Pepsin A (porcine residue 60-385) is an acidic protease (p*I* of 3.2) and a phosphoprotein which has one phosphoserine residue (Kinoshita et al., 2006). An optimum pH of pepsin A for its protease activity is around 2.0, meaning that pepsin A does not have any activity under our digestion condition at pH 8.5. Therefore, pepsin A was used as a substrate without any denaturation procedure. Similar to the digestion of β-casein, pepsin A was efficiently digested by CT-microreactor but not by TY-microreactor thus, explaining the difference between the total number of cleavage sites by CT (65 residues) and those by TY (3 residues). The sequence coverage of 55% (179/326 amino acids) by CT-microreactor was lower than that by in-solution digestion (60%, 196/326 amino acids) (Yamaguchi et al., 2010a). In addition, the phosphopeptide (EATpSQELSITY) was detected in the digestion by in-solution but not by CT-microreactor. When HPLC profile of the digests by the single CT-microreactor was compared with CT-AP tandem microreactor digest, it was found out that one peak disappeared after passing through the tandem microreactor. This suggests that the digests by CT-microreactor also contain the phosphopeptide but the size of the peptide was bigger than the EATpSQELSITY phosphopeptide from in-solution digestion. In addition, it is possible that our MS system was not able to detect the phosphopeptide from CT-microreactor. A possible reason for the lower digestion ability of CT-microreactor could be mass transfer limitation of folded pepsin A in the cross-linked CT and poly-Lys complex matrix. Similar lower digestion ability of the immobilized-protease was observed in the digestion of Cyt-C by CT-microreactor as described above.

The present analytical method of protein phosphorylation is much simpler than the other conventional methods (Han et al., 2009; Kinoshita et al., 2006; Zhao & Jensen, 2009), for example, the phosphoprotein is just flowed through the microreactor and it eliminates purification of digests from the reaction system without any enrichment strategies. These interesting features are superior advantages of our approach using the enzyme-immobilized microreactors over the conventional method.

### 3.6 Analysis of disulfide bond using the protease-immobilized microreactor at several temperatures

Disulfide bond is covalent cross-linking between side-chains of two Cys residues of the same or different peptide chains. It is an important factor for stabilizing the protein conformation. In addition, oxidation-reduction of Cys residues in protein is significant to biological functions (Lee et al., 2004; Mieyal et al., 2008; Yano & Kuroda, 2008). Therefore, the assignment of disulfide bonding patterns will not only provide insights into its three-dimensional structure and contribute to the understanding of its structure-function relationship but will also play a role in the regulation of fundamental cellular processes. However, the conventional method for the assignment of disulfide bond by chemical cleavage and/or proteolysis is a time-consuming multi-step procedure. In addition, due to higher conformational stability of protein by disulfide bond(s), the conventional in-solution digestion by protease was usually carried out using reduced protein which was prepared by the reduction and alkylation procedure. Although this approach has provided us with information on the protein sequence (primary structure of protein), the information for disulfide bond such as Cys-Cys pair(s), the number of disulfide bond, and a distinction between Cys residues involving disulfide bond and free Cys residue are not from the conventional reduced sample.

As shown above, the protease-immobilized microreactor showed a rapid and efficient digestion compared with in-solution digestion. Furthermore, the immobilized-protease can be easily isolated and removed from the digests prior to MS without any purification step. Moreover, the immobilized proteases by our methods were more stable at high temperature than free proteases (Yamaguchi et al., 2009). Based on these interesting features, we tested whether the substrate proteins that maintain their disulfide bonds can be efficiently digested by the protease-immobilized microreactor at high temperature without any chemical modification and purification step, and whether the position of disulfide bond(s) in the resultant digests can be analyzed by MS (Yamaguchi et al., 2010b).

It was reported that thermally denatured proteins were efficiently digested by in-solution digestion (Park & Russell, 2000) or by protease-microreactors using free proteases (Liu et al., 2009; Sim et al., 2006). Our approach is based on the concept that proteins of stable conformation owing to their disulfide bonds(s) were thermally denatured at high temperature and were directly digested by the protease-immobilized microreactor. It is known that thermal denaturation at higher temperature (~ 90 °C) would form the protein aggregation (Park & Russell, 2000) although the mechanism of this formation has not yet been fully elucidated. Therefore, we performed the proteolysis between 30 to 50 °C. Substrate proteins (50 µg/ml) which were not treated with the reduction and alkylation procedure were pumped through the microreactor at flow rate of 1.2 µl/min (reaction time: 21.7 min). The digests were collected in a test-tube and directly analyzed by ESI-TOF MS without any purification or concentration procedure.

Lysozyme (chicken residue 19-147: MW 14,304 Da) has well-characterized four disulfide bonds. As shown in Fig. 3, the digests by both microreactors showed that increasing reaction temperature increased the number of digested fragments, which were correlated with the sequence coverage (Table 2). Most of the digested peptides by the microreactors were the expected peptides which did not have the missed cleavage site. At 50 °C by TY-microreactor, 10 peptides containing 126 out of the 129 possible amino acids of lysozyme were obtained, producing the sequence coverage of 98%. This value was higher than those by TY-microreactor at 30 °C (5%, 7/129 amino acids) and in-solution digestion at 37 °C (77%, 99/129 amino acids). Moreover, all four disulfide bonds (Cys24-Cys145, Cys48-Cys133, Cys83-Cys98, and Cys94-Cys112) were detected in MS spectrum from the digests at 50 °C but 3 of 4 disulfide bonds (Cys48-Cys133, Cys83-Cys98, and Cys94-Cys112) were from in-solution digestion at 37 °C (Yamaguchi et al., 2010b). The sequence coverage of 22% (28/129) by the TY-microreactor at 40 °C was lower than that by in-solution digestion at 37 °C. A possible reason for the lower digestion ability of TY-microreactor could be that the substrate protein has different exposure of the digestion site by TY between in the microchannel and in batch method (in-solution digestion).

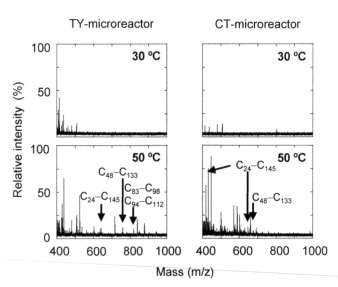

Fig. 3. ESI-TOF MS spectra of the digests of lysozyme by TY- or CT-microreactor. Digestion was carried out at different temperature: 30 and 50 °C. The peaks of disulfide bond(s)-containing peptides are marked with solid arrows.

| Digestion methods | Temperature / °C | Reaction time | Identified amino acids | Sequence coverage / % |
|---|---|---|---|---|
| TY-microreactor | 30 | 21.7 min | 7/129 | 5 |
| TY-microreactor | 50 | 21.7 min | 126/129 | 98 |
| CT-microreactor | 30 | 21.7 min | 11/129 | 9 |
| CT-microreactor | 50 | 21.7 min | 54/129 | 42 |
| TY (in-solution)[a] | 37 | 18 hours | 99/129 | 77 |
| CT (in-solution)[a] | 37 | 18 hours | 66/129 | 51 |

[a]In-solution digestion was carried out in 10 mM ammonium acetate buffer, pH 8.5. Concentrations of substrate and free proteases were 100 and 2 µg/ml, respectively.

Table 2. Summary of ESI-TOF MS results of the digests of lysozyme

In contrast, sequence coverage by CT-microreactor at 50 °C and in-solution digestion at 37 °C by free CT were lower than that by TY-microreactor (Table 2), although the total number of Phe, Trp, Tyr, Leu, and Met residues (22) for CT was larger than that of Arg and Lys residues (17) for TY. CD and fluorescence spectra measurements indicated that lysozyme at 50 °C was thermally denatured but partly formed some secondary structures. Basic residues are usually located on the surface of protein than hydrophobic residues. In addition, the p$I$ value of lysozyme is 9.3. Therefore, it is possible that TY easily recognizes and hydrolyses peptide bonds after Arg and Lys residues that locate on the surface of lysozyme. Because sequence coverage by CT was lower than that by TY, the number of identified disulfide bonds was also low. The sequence coverage of lysozyme by in-solution digestion at 50 °C (18% for free TY and 22% for free CT) were lower than those by the microreactors at 50 °C, indicating that free proteases decreased their hydrolysis activities at 50 °C.

To further investigate the efficiency of our approach for the assignment of disulfide bond, the digestion of bovine serum albumin (BSA) by the microreactor was carried out (Yamaguchi et al., 2010b). BSA (bovine residue 25-607: MW 66,390 Da) has 17 disulfide bonds. Similar to the digestion of lysozyme, the digestion of BSA by TY-microreactor at 50 °C efficiently occurred but not at 30 °C. The sequence coverage of 37% (201/583 amino acids) was higher compared with that of in-solution digestion (26%, 151/583 amino acids) and was better or comparable to those of a thermal (Liu et al., 2009; Sim et al., 2006) or chemical denatured BSA (Li et al., 2007a; Chatterjee et al., 2010) or the reduced BSA (Ma et al., 2008) by the reported trypsin-microreactors (24-46%). In addition, the number of identified disulfide bonds was 10 of 17, which is superior to 6 obtained by in-solution digestion. The digests of BSA by the TY-microreactor at 50 °C was also analyzed by MALDI-TOF MS. In addition to the information on disulfide bonds from ESI-TOF MS, we identified the TCVADESHAGCEK peptide with intramolecular disulfide bond (Cys77-Cys86) that was not identified by ESI-TOF MS. Although all disulfide bonds in BSA were not identified under the present condition, these results indicate that proteolysis approach at high temperature by the microreactors can be useful not only for large proteins but also for the proteins having many disulfide bonds. After each proteolysis procedure at several temperatures, the microreactors were washed with buffer solution and digestions were carried out at 30 °C. The residual activities of both microreactors after proteolysis procedure even at 50 °C were identical with those of prior to proteolysis, indicating that both immobilized proteases maintained their activities after several use at different temperatures. We also tested the hydrolysis activity of TY-microreactor against Cyt-C. MS spectra of Cyt-C by TY-microreactor before and after proteolysis at 50 °C were the same. In addition, both sequence coverage of Cyt-C by TY-microreactor were 89% (93/104 amino acids). This value was higher than those of the reported trypsin-immobilized reactors (Liu et al., 2007a; Liu et al., 2009). In contrast, free proteases showed lower proteolysis activities at 50 °C than those at 37 °C. Once again, these results verified that the present protease-immobilized microreactors were thermally stable at high temperature but this was not observed in the case of free proteases. Similar stability of immobilized TY on polymer nanofibers was previously reported (Kim et al., 2009).

From these results, the procedures of proteolysis by the microreactor and the MS measurement took less than 2 hours. This is much faster and easier than the conventional procedure (multi-days). Therefore, our proteolysis approach by the thermostable microreactor is a simple and rapid analytical method for the assignment of disulfide bond without any chemical modification or purification procedure.

## 4. Conclusion

The microreactors showed efficient proteolysis with high sequence coverage, long-term stability, and good reusability. It is known that in-solution digestion by trypsin can induce artificial modifications such as asparagine deamidation (Krokhin et al., 2006) and N-terminal glutamine cyclization (Bongers et al., 1992; Dick et al., 2007) on target protein due to the elevated temperature and alkaline pH buffers used during digestion for overnight. Proteolysis by our protease-immobilized microreactors was achieved within a short period of time (~ 20 min) at 30 °C therefore suggesting these artificial modifications as a remote possibility. In addtion, proteolysis by the tandem microreactors showed higher sequence coverage, which is a remarkable result compared with those of the single microreactor or in-solution digestion. The tandem microreactor comprising a protease-microreactor and a phosphatase-microreactor also showed the capability to localize phosphorylation site(s) in phosphoproteins. Several protease-immobilized microreactors were developed for proteolysis. Most of these studies have focused on rapid digestion and reduction of sample volume. So far, there is no study yet on multi-enzymatic reaction system and analysis of post-translational modification in protein. Furthermore, proteolysis at 50 °C by the microreactors showed higher sequence coverage and assignment of disulfide bonds, which is a remarkable result compared with that of in-solution digestion.

The present procedure is much simpler than the other conventional methods, for example, the protein is just flowed through the microreactor and it is not necessary to purify the digests from the reaction system. These interesting features are superior advantages of our proteolysis approach over the conventional method. The enzyme-immobilization method using poly-Lys can be applied to proteins with wide-range p*I* values hence, the strategy based on multi-enzymatic reaction using the tandem microreactor provides a useful approach for other post-translational modification analysis (*e.g.* acetylation, methylation, ubiquitination or glycosylation). Coupling the protease-immibilized microreactor with MS and/or HPLC (on-line) can be also applied for high throughput proteomic analysis systems.

## 5. Acknowledgment

This work was supported by Grant-in-Aid for Basic Scientific Research (B: 23310092) and for Young Scientists (B: 23710153), from the Japan Society for the Promotion of Science (JSPS).

# 6. References

Aebersold, R. & Mann, M. (2003) Mass spectrometry-based proteomics. *Nature*, 422, 198-207.

Asanomi, Y., Yamaguchi, H., Miyazaki M. & Maeda, H. (2011) Enzyme-immobilized microfluidic process reactors. *Molecules*, 16, 6041-6059.

Bongers, J., Heimer, E. P., Lambros, T., Pan, Y. C., Campbell, R. M. & Felix, A. M. (1992) Degradation of aspartic acid and asparagines residues in human growth hormone-releasing factor. *Int. J. Pept. Protein Res.*, 39, 364-374.

Chatterjee, D., Ytterberg, A. J., Son, S. U., Loo, J. A. & Garrell, R. L. (2010) Integration of protein processing steps on a droplet microfluidics platform for MALDI-MS analysis. *Anal. Chem.*, 82, 2095-2101.

Chen, W. Y. & Chen Y. C. (2007) Acceleration of microwave-assisted enzymatic digestion reactions by magnetite beads. *Anal. Chem.*, 79, 2394-2401.

Dick Jr., L. W., Kim, C., Qiu, D. & Cheng, K. C. (2007) Determination of the origin of the N-terminal pyro-glutamate variation in monoclonal antibodies using model peptides. *Biotechnol. Bioeng.* 97, 544-553.

Domon, B. & Aebersold, R. (2006) Mass spectrometry and protein analysis. *Science*, 312, 212-217.

Dulay, M. T., Baca, Q. J. & Zare, R. N. (2005) Enhanced proteolytic activity of covalently bound enzymes in photopolymerized sol gel . *Anal. Chem.*, 77, 4604-4610.

Ethier M., Hou, W., Duewel, H. S. & Figeys, D. (2006) The proteomic reactor: a microfluidic device for processing minute amounts of protein prior to mass spectrometry analysis. *J. Proteomie Res.*, 5, 2754-2759.

Fan, H. & Chen, G. (2007) Fiber-packed channel bioreactor for microfluidic protein digestion. *Proteomics*, 7, 3445-3449.

Fischer, F. & Poetsch, A. (2006) Protein cleavage strategies for an improved analysis of the membrane proteome. *Proteome Sci*, 4, 2.

Han, G., Ye, M., Jiang, X., Chen, R., Ren, J., Xue, Y., Wang, F., Song, C., Yao, X. & Zou, H. (2009) Comprehensive and reliable phosphorylation site mapping of individual phosphoproteins by combination of multiple stage mass spectrometric analysis with a target-decoy database search. *Anal. Chem.*, 81, 5794-5805.

Honda, T., Miyazaki, M., Nakamura, H. & Maeda, H. (2005) Immobilization of enzymes on a microchannel surface through cross-linking polymerization. *Chem. Commun.*, 5062-5064.

Honda, T., Miyazaki, M., Nakamura, H. & Maeda, H. (2006) Facile preparation of an enzyme-immobilized microreactor using a cross-linking enzyme membrane on a microchannel surface. *Adv. Synth. Catal.* 348, 2163-2171.

Honda, T., Miyazaki, M., Yamaguchi, Y., Nakamura, H. & Maeda, H. (2007) Integrated microreaction system for optical resolution of racemic amino acids. *Lab. Chip.*, 7, 366-372.

Hunter, T. (2009) Tyrosine phosphorylation: thirty years and counting. *Curr. Opin. Cell Biol.*, 21, 140-146.

Kim, B. C., Lopez-Ferrer, D., Lee, S. -M., Ahn, H. -K., Nair, S., Kim, S. H., Kim, B. S., Petritis, K., Camp, D. G., Grate, J. W., Smith, R. D., Koo, Y. M., Gu, M. B. & Kim, J. (2009) Highly stable trypsin-aggregate coatings on polymer nanofibers for repeated protein digestion. *Proteomics*, 9, 1893-1900.

Kinoshita, E., Kinoshita-Kikuta, E., Takiyama, K. & Koike, T. (2006) Phosphate-binding tag, a new tool to visualize phosphorylated proteins. *Mol. Cell. Proteomics*, 5, 749-757.

Krokhin, O. V., Antonovici, M., Ens, W., Wilkins, J. A. & Standing, K. G. (2006) Deamidation of -Asn-Gly- sequences during sample preparation for proteomics: Consequence for MALDI and HPLC-MALDI analysis. *Anal. Chem.*, 78, 6645-6650.

Lee, K., Lee, J., Kim, Y., Bae, D., Kang, K. Y., Yoon, S. C. & Lim, D. (2004) Defining the plant disulfide proteome. *Electrophoresis*, 25, 532-541.

Lee, J., Musyimi, H. K., Soper, S. A. & Murray, K. K. (2008) Development of an automated digestion and droplet deposition microfluidic chip for MALDI-TOF MS. *J. Am. Soc. Mass Spectrom.*, 19, 964-972.

Li, Y., Xu, X. Q., Deng, C. H., Yang, P. Y. & Zhang, X. M. (2007a) Immobilization of trypsin on superparamagnetic nanoparticles for rapid and effective proteolysis. *J. Proteome Res.*, 6, 3849-3855.

Li, Y., Xu, X., Qi, D., Deng, C., Yang, P. & Zhang, X. (2007b) Novel $Fe_3O_4$ @ $TiO_2$ core-shell microspheres for selective enrichment of phosphopeptides in phosphoproteome analysis. *J. Proteome Res.*, 7, 2526-2538.

Lin, S., Yao G., Qi, D., Li, Y., Deng, C., Yang, P. & Zhang X. (2008) Fast and efficient proteolysis by microwave-assisted protein digestion using trypsin-immobilized magnetic silica microspheres. *Anal. Chem.*, 80, 3655-3665.

Liu, Y., Lu, H., Zhong, W., Song, P., Kong, J., Yang, P., Girault, H. H. & Liu B. (2006) *Anal. Chem.*, 78, 801-808.

Liu, Y., Liu, B., Yang, P. & Girault, H. H., (2008) Microfluidic enzymatic reactors for proteome research. *Anal. Bioanal. Chem.*, 390, 227-229.

Liu, T., Bao, H., Zhang, L. & Chen, G. (2009) Integration of electrodes in a suction cup-driven microchip for alternating current-accelerated proteolysis. *Electrophoresis*, 30, 3265-3268.

Ma, J., Ziang, Z., Qiao, X., Deng, Q., Tao, D., Zhang L. & Zhang, Y. (2008) Organic-inorganic hybrid silica monolith based immobilized trypsin reactor with high enzymatic activity. *Anal. Chem.*, 80, 2949-2956.

Ma, J., Zhang, L., Liang, Z., Zhang, W. & Zhang, Y. (2009) Recent advance in immobilized enzymatic reactors and their applications in proteome analysis. *Anal. Chim. Acta*, 632, 1-8.

Mieyal, J. J., Gallogly, M. M., Qanungo, S., Sabens, E. A. & Shelton, M. D. (2008) Molecular mechanisms and clinical implications of reversible protein *S*-glutathionylation. *Antioxid. Redox Signal.*, 10, 1941-1988.

Miyazaki M. & Maeda H. (2006) Microchannel enzyme reactors and their applications for processing. *Trends Biotechnol.*, 24, 463-470.

Miyazaki, M., Honda, T., Yamaguchi, H., Briones, M. P. P. & Maeda, H. (2008) Enzymatic processing in microfluidic reactors., *Biotechnol. Genet. Eng. Rev.*, 25, 405-428.

Nel, A. L., Krenkova, J., Kleparnik, K., Smadja, C., Taverna, M., Viovy, J. L. & Foret, F. (2008) On-chip tryptic digest with direct coupling to ESI-MS using magnetic particles. *Electrophoresis*, 29, 4944-4947.

Park Z. Y. & Russell D. H. (2000) Thermal denaturation: A useful technique in peptide mass mapping. *Anal. Chem.*, 72, 2667-2670.

Sakai-Kato, K., Kato, M. & Toyooka, T. (2002) On-line trypsin-encapsulated enzyme reactor by the sol-gel method integrated into capillary electrophoresis. *Anal. Chem.*, 74, 2943-2949.

Sakai-Kato, K., Kato M. & Toyooka, T. (2003) Creation of an on-chip enzyme reactor by encapsulating trypsin in sol-gel on a plastic microchip. *Anal. Chem.*, 75, 388-393.

Sheldon R. A. (2007) Enzyme immobilization: The quest for optimum performance. *Adv. Synth. Catal.* 349, 1289-1307.

Sim, T. S., Kim, E. -M., Joo, H. S., Kim, B. G. & Kim, Y. -K. (2006) Application of a temperature-controllable microreactor to simple and rapid protein identification using MALDI-TOF MS. *Lab. Chip*, 6, 1056-1061.

Slysz, G. W., Lewis, D. F. & Schriemer, D. C. (2006) Detection and identification of sub-nanogram levels of protein in nanoLC-trypsin-MS system. *J. Proteome Res.*, 5, 1959-1966.

Shui, W., Fan, J, Yang, P., Liu, C., Zhai, J., Lei, J., Yan, Y., Zhao, D. & Chen, X. (2006) Nanopore-based proteolytic reactor for sensitive and comprehensive proteomic analyses. *Anal. Chem.*, 78, 4811-4819.

Temporini, C., Calleri, E., Cabrera, K., Felix, G. & Massolini, G. (2009) On-line multi-enzymatic approach for improved sequence coverage in protein analysis. *J. Sep. Sci.* 32, 1120-1128.

Wang, L. S., Khan, F. & Micklefield, J. (2009) Selective covalent protein immobilization: strategies and applications. *Chem. Rev.*, 109, 4025-4053.

Witze, E. S., Old, W. N., Resing, K. A. & Ahn, N. G. (2007) Mapping protein post-translational modifications with mass spectrometry. *Nat. Methods*, 10, 798-806.

Wu H., Tian Y., Liu, B., Lu, H., Wang, X., Zhai, J., Jin, H., Yang, P., Xu, Y. & Wang H. (2004) Titania and alumina sol−gel-derived microfluidics enzymatic-reactors for peptide mapping: design, characterization, and performance. *J. Proteome Res.*, 3, 1201-1209.

Yamaguchi, H., Miyazaki, M., Honda, T., Briones-Nagata, M. P., Arima, K. & Maeda, H. (2009) Rapid and efficient proteolysis for proteomic analysis by protease-immobilized microreactor. *Electrophoresis*, 30, 3257-3264.

Yamaguchi, H., Miyazaki, M., Kawazumi, H. & Maeda, H. (2010a) Multidigestion in continuous flow tandem protease-immobilized microreactors for proteomic analysis. *Anal. Biochem.*, 407, 12-18.

Yamaguchi, H., Miyazaki, M. & Maeda, H. (2010b) Proteolysis approach without chemical modification for a simple and rapid analysis of disulfide bonds using thermostable protease-immobilized microreactors. *Proteomics*, 10, 2942-2949.

Yamaguchi, H., Miyazaki, M., Asanomi Y. & Maeda, H. (2011) Poly-lysine supported cross-linked enzyme aggregates with efficient enzymatic activity and high operational stability. *Catal. Sci. Technol.*, DOI: 10.1039/c1cy00084e.

Yamashita, K., Miyazaki, M., Nakamura, H. & Maeda, H. (2009) Nonimmobilized enzyme kinetics that rely on laminar flow. *J. Phys. Chem. A*, 113, 165-169.

Yano, H. & Kuroda, S. (2008) Introduction of the disulfide proteome: Application of a technique for the analysis of plant storage proteins as well as allergens. *J. Proteome Res.*, 7, 3071-3079.

Zhao, Y. & Jensen, O. N. (2009) Modification-specific proteomics: strategies for characterization of post-translational modifications using enrichment techniques. *Proteomics*, 9, 4632-4641.

# 6

# Labeling Methods in Mass Spectrometry Based Quantitative Proteomics

Karen A. Sap and Jeroen A. A. Demmers
*Erasmus University Medical Center*
*The Netherlands*

## 1. Introduction

Proteomics is loosely defined as the description of sets of proteins from any biological source, which have in most cases been identified by using mass spectrometry. However, only the mere identity of proteins present in a certain sample does not give any information about the dynamics of the proteome, involving relevant cellular events such as protein synthesis and degradation, or the formation of protein assemblies. In order to retrieve information on proteome dynamics, relative protein abundances between different protein samples should be assessed. Comparative or differential proteomics aims to identify *and* quantify proteins in different samples, to study *e.g.* differences between healthy and diseased states, mutant and wildtype cell lines, undifferentiated and differentiated cells, etc. Since mass spectrometry is in itself only a qualitative technique, various methods to obtain quantitative information of the proteome have been developed over the past decade and will be described in this Chapter.

We will focus on post-digestion labeling methods in the field of functional proteomics. Functional proteomics focuses on characterizing the composition of protein complexes, and generally involves the affinity purification of a protein of interest followed by the identification of co-purifying proteins by mass spectrometry (AP-MS). Generally, proteins in a negative control sample and those identified in the sample containing the protein of interest and its interacting partners are directly compared to determine which of the proteins interact in a specific manner. However, the mere presence or absence of a certain protein in protein data sets as a measure for either overlap or specificity is generally not sufficient, as this gives no information about the relative abundances of the present proteins. A generally recognized problem is the presence of contaminating proteins that are identified in the mass spectrometric screen, but do not really make part of the protein complex. Often, these background proteins are highly abundant proteins that stick to the complex or to beads to which the antibody is conjugated in a non-specific manner. A more accurate and correct approach would therefore involve a strategy in which protein abundance differences between sample and control can be assessed in a quantitative manner and which helps in discriminating *bona fide* interaction partners from such background proteins. Ideally, a differential mass spectrometric method would allow for an unbiased, sensitive, and high-throughput screening for protein-protein interaction networks.

## 1.1 SDS-PAGE based methods for protein quantitation

Two-dimensional sodium dodecylsulfate polyacrylamide gel electrophoresis (2D-SDS-PAGE) has traditionally been a popular method for differential-display proteomics on a global scale, although recently the popularity and applicability of stable isotope LC-MS based methods has exceeded those presented by gel based methods. 2D-SDS-PAGE based methods enable the separation of complex protein mixtures on a single gel. Proteins are separated in two dimensions: in the first dimension, they are fractionated according to their isoelectric point using a pH gradient gel, which is subsequently placed on a polyacrylamide gel slab for further separation based on their molecular weight using SDS-PAGE. Proteins are then visualized by staining the gel with a dye such as Coomassie, silver or Sypro Ruby. In principle, for comparative issues, samples are loaded on separate gels and protein spot patterns are compared visually. Proteins that differ in abundance can then be punched out of the gel, digested with a suitable protease and analyzed by mass spectrometry. In a variation of this technique, difference gel electrophoresis (DIGE), proteins from two samples are first labeled with different fluorescent dyes and then mixed, making it possible to compare two different samples on a single gel. Two fluorescence images are recorded and overlayed, and differentially expressed proteins appear in only one of the images (Unlu et al., 1997). Limitations of this method include the manual selection of proteins to be analyzed, making it a time-consuming technique, as well as the limited sensitivity, as a consequence of which that proteins with a low concentration may be failed to be selected. Nowadays, in many laboratories there is a tendency to replace 2D-SDS-PAGE based methods by more powerful, LC-MS based methods for relative protein quantitation.

## 1.2 Protein and peptide quantitation using LC-MS based methods

Rather than by comparing protein spot intensities on a gel, quantification of proteins in LC/MS based methods is based on the peak height or area of the proteolytic peptide peaks in the mass spectrum and/or chromatogram. As mentioned before, mass spectrometry is not an inherently quantitative analytical technique, meaning that the peak height or area in a mass spectrum in itself does not accurately reflect the abundance of a peptide in the sample. The main reasons for this are the differences in ionization efficiency and detectability of peptides because of their different physicochemical characteristics, as well as the limited reproducibility of an LC-MS experiment. Altogether, this makes it difficult to compare peptide peak intensities between different mass spec runs. In principle however, peak intensity differences of the same analyte within one LC-MS run do accurately reflect the abundance difference. One way to distinguish the same analyte from different sample sources within one LC-MS experiment is by using stable heavy isotope labeling. When different stable isotope labels are used for proteins or peptides which are derived from different samples, the same analyte can in principle be quantified in one experiment. Such heavy stable isotope labels should in principle not affect the biophysical and chemical properties of peptides and proteins, but solely the mass, designating one of the samples as 'light' and the other sample 'heavy' according to the mass introduced by the label. The heavy and light peptides co-elute from the LC column at the same retention time and the heavy stable isotope leads to a mass shift in the mass spectrum, resulting in the observation of peak pairs. The peak heights or areas of such pairs can be compared and give an accurate reflection of the difference in abundance of this peptide between both samples. Heavy stable isotope labels can be introduced at different stages in the sample treatment protocol. Below, we will give an overview of the most widely used labeling techniques.

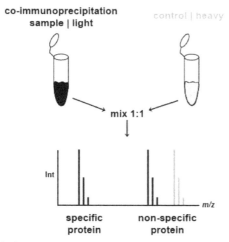

Fig. 1. In a differential labeling AP-MS experiment, proteins in a control sample are labeled with a heavy stable isotope label, whereas proteins in the experimental sample are labeled with a light label. Incorporation of the heavy label results in a shift of the *m/z* value and allows one to differentiate between the sources of the protein of interest.

## 2. LC-MS-based quantitation methods

### 2.1 Incorporation of stable isotopes by metabolic labeling

Heavy stable isotope labels can be introduced *in vivo* by growing cells or even whole organisms in the presence of amino acids or nutrients carrying such stable isotopes. Metabolic labeling is often the preferred labeling technique, since incorporation occurs at the earliest possible moment in the sample preparation process, thereby minimizing the error in quantification (see Figure 2). Several methods based on metabolic labeling have been developed and here we will give a brief overview. The first metabolic labeling studies were performed utilizing [15]N-enriched media to grow *S. cerevisiae* (Oda et al., 1999) and *E. coli* (Conrads et al., 2001). Next, the method was extended towards multicellular organisms which were [15]N labeled, such as *D. melanogaster* and *C. elegans* by feeding them on labeled yeast or bacteria, respectively (Krijgsveld et al., 2003). Even a higher eukaryote like a rat has been labeled with [15]N (McClatchy et al., 2007). Plants, as they are autotrophic organisms, can easily be labeled metabolically through feeding of labeled inorganic compounds in the form of [15]N-nitrogen-containing salts, as first demonstrated in NMR studies (Ippel et al., 2004), and later in MS-based proteomics (Engelsberger et al., 2006; Lanquar et al., 2007). [15]N atoms are incorporated into the sample during cell growth, eventually replacing all natural isotopic (*i.e.*, [14]N) nitrogen atoms. The corresponding mass shift depends on the number of nitrogen atoms present in each of the resulting proteolytic peptides. However, this variable mass shift complicates data analysis to a large extent and requires high resolution mass spectrometry for the analysis (Conrads et al., 2001). Specific software for the analysis of [15]N labeled samples has been developed (Mortensen et al., 2010).

Stable isotope labeling in cell culture (SILAC) is a metabolic labeling approach first published in 2002 by the lab of Matthias Mann (Ong et al., 2002). During cell growth, essential amino acids that carry heavy stable isotopes and which have been added to the

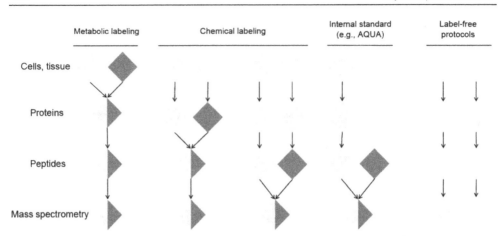

Fig. 2. Stages of incorporation of stable isotope labels in typical labeling workflows in quantitative proteomics. The light and dark grey diamonds represent the two protein samples to be differentially labeled and compared. Figure adapted from (Ong & Mann, 2005).

culturing medium are introduced in all newly synthesized proteins. After several cell doublings, the complete cellular proteome will have incorporated the supplied labeled amino acid(s). This results in a shift of the proteolytic peptide mass after protein digestion and subsequent MS analysis. When labeled and non-labeled cell cultures are now mixed and analyzed in the same experiment, peptides will be represented by peak pairs in the mass spectrum, where the mass difference will depend on the number and nature of the labeled amino acid(s). Usually, labeled lysine and arginine are used, with the result that every peptide will carry a label except for the carboxyl-terminal peptide of the protein, when digested with trypsin, as does labeling with lysine when digested with Lys-C (Ibarrola et al., 2003). In contrast to $^{15}N$ labeling, the number of incorporated labels in SILAC is defined and not dependent on the peptide sequence, thus facilitating data analysis. SILAC has been successfully applied in global proteome studies (de Godoy et al., 2006), for functional proteomics assays, as well as for the study of post-translational modifications (Blagoev et al., 2003; Blagoev & Mann, 2006).

Because of the label incorporation at early stages in the sample preparation protocol, SILAC is generally the preferred choice of labeling method. However, SILAC is limited in sample applicability, for example, not every cell line can grow in an efficient manner in media optimized for SILAC, often due to the requirement of dialyzed serum in the medium to prevent contamination with natural amino acids. Besides, the method may be hampered by *in vitro* conversion of labeled arginine to proline (Van Hoof et al., 2007). SILAC has been used to label higher organisms, for instance flies (Sury et al., 2010) and mice (Kruger et al., 2008), by feeding them with labeled food. In general though, this is a time consuming and expensive process. In the plant, SILAC has only yielded label incorporation of approximately 70% (Gruhler et al., 2005), which is not satisfying for many proteomics applications. Moreover, there are practical and moral limitations to SILAC labeling of human tissue. For these cases, methods for stable isotope label incorporation at a later stage in the sample preparation protocol are required. Chemical and enzymatic labeling techniques have been developed that can introduce the heavy stable isotope label only after sample collection and proteolytic digestion at the peptide level.

## 2.2 Incorporation of stable heavy isotope labels by chemical or enzymatic labeling

In general, the advantage of chemical labeling over metabolic labeling is the possibility to label a wide range of different sample types, since incorporation of the label is performed only after harvesting cells and subsequent purification of proteins. Chemical labeling is essentially based on similar mechanisms as metabolic labeling, except that the label is introduced into proteins or peptides by a chemical reaction, *e.g.*, with sulfhydryl groups or amine groups, or through acetylation or esterfication of amino acid residues. Alternatively, the heavy stable isotope label can be introduced into the peptide during an enzymatic reaction with heavy water ($H_2^{18}O$). Below, several of the most widely applied chemical and enzymatic labeling approaches are described.

### 2.2.1 Isotope-Coded Affinity Tags (ICAT)

Isotope-Coded Affinity Tagging (ICAT) is a chemical labeling method that was first described by the Aebersold lab in 1999 (Gygi et al., 1999). In chemical modification-based approaches, stable isotope-bearing chemical reagents are targeted towards reactive sites on a protein or peptide. The ICAT reagent consists of a reactive group that is cysteine-directed, a polyether linker region with eight deuteriums, and a biotin group that allows purification of labeled peptides. In an ICAT experiment, two pools of proteins are denatured and reduced, and the cysteine residues of the proteins are subsequently derivatized with either the 'heavy' or 'light' ICAT reagent. The labeled pools are then combined, cleaned up to remove excess reagent, and digested with an appropriate protease. The cysteine-containing peptides, carrying 'heavy' and 'light' isotope tags, are then captured on an avidin column via the biotin moiety present at the incorporated label. Peptides are then eluted from the column and analyzed by mass spectrometry. Since only cysteine-containing peptides are isolated, the peptide mixture complexity is in general limited, which in principle would enable identification of lower abundant proteins. On the other hand, some proteins contain no cysteines, while others would have to be quantified on the basis of just a single peptide. Additionally, the large biotin tag significantly increases the complexity of fragmentation spectra, complicating peptide identification, and, besides that, it has been demonstrated that deuterium atoms that are associated with the tag can cause a shift in retention time between the light and heavy peptides in reverse phase chromatography (Zhang et al., 2001). Subsequent iterations of the ICAT approach by substituting a cleavable and co-eluting tag have improved the method (Hansen et al., 2003; Li et al., 2003).

### 2.2.2 Dimethyl labeling

An alternative method based on chemical labeling is dimethylation of peptides. In this workflow, samples are first digested with proteases such as trypsin and the derived peptides of the different samples are then labeled with isotopomeric dimethyl labels. The labeled samples are mixed and simultaneously analyzed by LC-MS whereby the mass difference of the dimethyl labels is used to compare the peptide abundance in the different samples. Stable isotope labeling by dimethylation is based on the reaction of peptide primary amines (peptide N-termini and the epsilon amino group of lysine residues) with formaldehyde to generate a Schiff base that is rapidly reduced by the addition of cyanoborohydride to the mixture. These reactions occur optimally between pH 5 and 8.5. Dimethyl labeling can be used as a triplex reagent, making it possible to quantitatively analyze three different samples in a single MS run. Labeling with the light reagent generates a mass increase of 28 Da per primary amine on a peptide and is obtained by using regular formaldehyde and cyanoborohydride. Using deuterated formaldehyde in combination with regular cyanoborohydride generates a mass increase of 32 Da per primary amine; this is referred to as the intermediate label (Hsu et al., 2003) . Incorporation of the heavy label can be achieved through combining deuterated and $^{13}$C-labeled formaldehyde with cyanoborodeuteride, resulting in a mass increase of 36 Da (Boersema et al., 2008). These reactions are visualized in Figure 3.

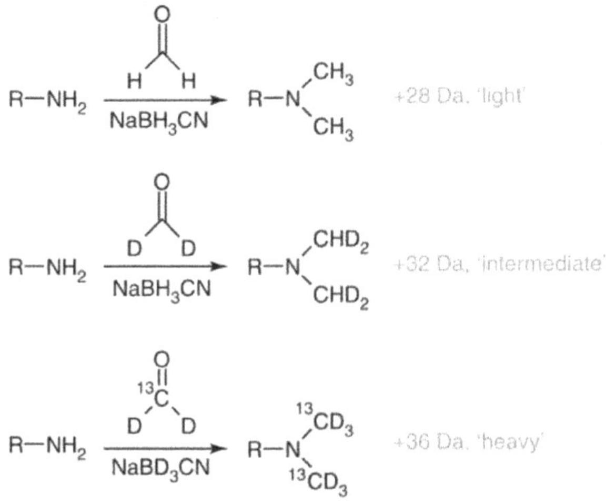

Fig. 3. Labeling schemes of triplex stable isotope dimethyl labeling. R: remainder of the peptide. Figure adapted from (Boersema et al., 2008).

One drawback of the incorporation of deuterium is that deuterated peptides show a small but significant retention time difference in reversed phase chromatography compared to their non-deuterated counterparts (Zhang et al., 2001). This complicates data analysis because the relative quantities of the two peptide species cannot be determined accurately from one spectrum but requires integration across the chromatographic time scale. As the stable isotope dimethyl labeling is performed at the peptide level, the method is not subjected to restrictions on the origin of the biological sample. Stable isotope dimethyl labeling can be performed in up to 8M urea, as well as after in-gel digestion protocols. It should be noted that during the sample preparation workflow, no buffers and solutions containing primary amines (such as ammonium bicarbonate and Tris) ought to be used, as formaldehyde would react with these, which would affect the labeling efficiency. This can be circumvented by desalting the peptide sample before the labeling reaction or by performing the digestion in buffers without primary amines (e.g., triethyl ammonium bicarbonate (TEAB)). Since both the peptide N-termini and lysine side chain amino groups are labeled in this protocol, it is compatible with the peptide products of virtually any protease, such as trypsin, Lys-C, Lys-N, Arg-C, and V8 (Boersema et al., 2008). Typically, for proteomics experiments trypsin is used, which cleaves C-terminal of lysine and arginine residues. Labeling of tryptic peptides using the method described here results in a mass shift of either 4 Da (when cleaved after an arginine residue) or 8 Da (when cleaved after a lysine residue) between the light and intermediate and between the intermediate and heavy label.

Differential labeling of peptides resulting from digestion with Lys-C or Lys-N (cleaving respectively C- and N-terminal of lysine residues) will result in a mass difference of mainly 8 Da (both the N-terminus and the lysine residues are labeled), whereas peptide products from Arg-C and V8 will result in varying mass differences as the number of lysine residues per peptide will typically vary. After proteolytic digestion, the samples are labeled separately by incubation with $CH_2O$ and $NaBH_3CN$ (light), $CD_2O$ and $NaBH_3CN$ (intermediate) or $^{13}CD_2O$ and $NaBD_3CN$ (heavy).

Boersema and co-workers have described three different experimental protocols for dimethyl labeling, *i.e.* in-solution, online, and on-column (Boersema et al., 2009). In-solution labeling (Boersema et al., 2008; Hsu et al., 2003) can be used for sample amounts from 1 μg to several milligrams of sample and is most suitable for experiments in which large sample numbers have to be labeled since labeling can be performed in parallel here. Online stable isotope labeling is the optimal method for the labeling of small quantities (<< 1 μg) of sample, because the sample loss is diminished by combining sample clean-up and labeling and by performing LC-MS directly after labeling. Finally, the on-column stable isotope labeling method is most suited for larger (up to milligrams) sample amounts, as sample clean-up and labeling steps are combined and the quenching step is avoided. After labeling, the samples are mixed and analyzed by mass spectrometry. Finally, quantification is performed by comparing the signal intensities of the differentially labeled peptides (see section on Data Analysis).

Protein quantitation by dimethyl labeling has been applied in a variety of studies, *e.g.* for the investigation of tyrosine phosphorylation sites in Hela cells upon EGF stimulation (Boersema et al., 2010). Proteins in a HeLa cell extract were dimethyl labeled and subsequently enriched for phosphorylated-tyrosine-containing peptides using immunoaffinity assays. Several tens of unique phosphotyrosine peptides were found to be regulated by EGF, illustrating that such a targeted quantitative phosphoproteomics approach has the potential to study signaling events in detail. Furthermore, the method has been applied to unravel differences in composition between highly related protein complexes, such as tissue-specific bovine proteasomes (Raijmakers et al., 2008) and the yeast nuclear and cytoplasmic exosome protein complex (Synowsky et al., 2009).

In conclusion, dimethyl labeling is a reliable, cost-effective and undemanding procedure that can be easily automated and applied in high-throughput proteomics experiments. It is applicable to virtually any sample, including tissue samples derived from animals or humans and up to three samples can be analyzed simultaneously. Like other chemical labeling methods though, stable isotope dimethyl labeling is performed in one of the final steps of a typical proteomics workflow and is therefore more prone to errors in the quantitative analysis as compared to workflows in which the label is added at an earlier stage.

### 2.2.3 $^{18}$O labeling

$^{18}$O labeling relies on class-2 proteases, such as trypsin, to catalyze the exchange of two $^{16}$O atoms for two $^{18}$O atoms at the C-terminal carboxyl group of proteolytic peptides, resulting in a mass shift of 4 Da between differently labeled peptides, as illustrated in Figure 4.

Hydrolysis of a protein in $H_2^{18}O$ by a protease results in the incorporation of one $^{18}$O atom into the carboxyl terminus of each proteolytically generated peptide. This mechanism involves a nucleophilic attack by a solvent water molecule on the carbonyl carbon of the scissile peptide bond (reaction 1). Following this hydrolysis reaction, the protease

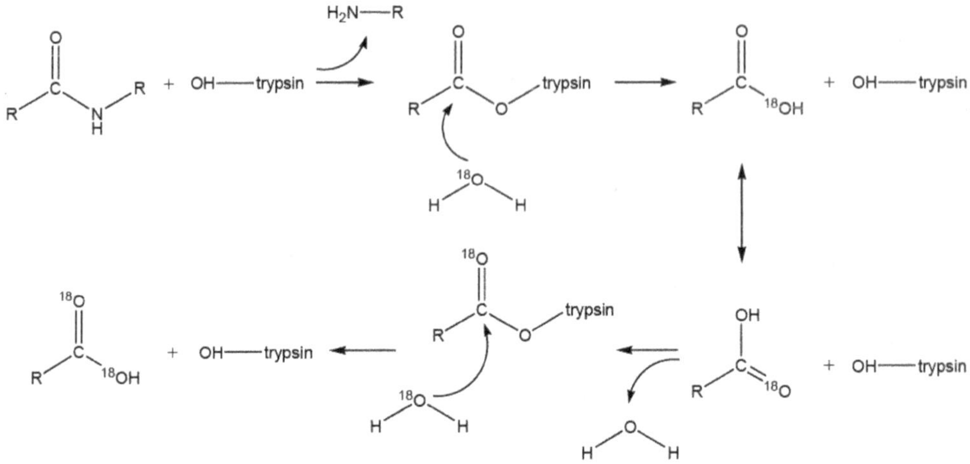

Fig. 4. Principle of trypsin catalyzed $^{18}O$ labeling. Incorporation of two $^{18}O$ labels at the C-terminus of a tryptic peptide takes place in a two-step reaction.

incorporates one more $^{18}O$ atom into the carboxyl terminus of the proteolytically generated peptide. This second incorporation results in two $^{18}O$ atoms being incorporated into the carboxyl terminus of the peptides (reaction 2). The second $^{18}O$ atom-incorporation is essentially the reverse reaction of peptide-bond hydrolysis or the peptide-bond formation reaction (Miyagi & Rao, 2007).

Proteolytic $^{18}O$ labeling has shown to be a useful tool in the field of comparative proteomics. A number of studies have been published, involving among others relative protein quantitation of the virus proteome (Yao et al., 2001), proteomes of cultured cells (Blonder et al., 2005; Brown & Fenselau, 2004; Rao et al., 2005) and proteins in serum (Hood, Lucas et al., 2005; Qian et al., 2005) and tissues (Hood, Darfler et al., 2005; Zang et al., 2004). In addition, $^{18}O$ labeling has been used for the relative quantitation of post-translational modification, e.g. changes of protein phosphorylation in response to a stimulus (Bonenfant et al., 2003). In the latter study, pools of differentially labeled phosphorylated proteins were enriched by using immobilized metal-affinity chromatography. Peptides were then dephosphorylated by alkaline phosphatase in order to quantify the changes in phosphorylation by mass spectrometry. A similar approach has been used for the global phosphoproteome analysis of human HepG2 cells (Gevaert et al., 2005).

Despite its relatively simple mechanism and low costs, $^{18}O$ labeling has not become the preferred method for differential proteomics based on heavy stable isotope labeling. The practical difficulties involved, most importantly the occurrence of incomplete incorporation of two $^{18}O$ atoms into the proteolytic peptide, and, as a consequence, the difficulties in data analysis and interpretation are the most likely reasons for this. Several factors are responsible for the variable degree of $^{18}O$ incorporation, including variable enzyme substrate specificity, oxygen back exchange, pH dependency and peptide physicochemical properties. To overcome inefficient labeling, algorithms for the correction of $^{18}O$ labeling efficiency have been developed (Ramos-Fernandez et al., 2007), while other studies have focused on minimizing back exchange of $^{18}O$ to $^{16}O$. It was found that the latter can be achieved by either decreasing the pH value for trypsin catalyzed incorporation reactions

(Hajkova et al., 2006; Staes et al., 2004; Zang et al., 2004), or by using immobilized trypsin for the exchange reaction (Chen et al., 2005; Fenselau & Yao, 2007; Sevinsky et al., 2007). Trypsin immobilization allows the investigator to significantly increase the molar ratio of protease-to-substrate ratio, which subsequently increases the labeling efficiency. Another advantage of using immobilized proteases is that no protease-catalyzed oxygen back exchange reaction occurs, because the immobilized proteases are completely removed from the peptides after the labeling reaction.

Our lab has developed a two-step approach in order to completely label all proteolytic peptides (Bezstarosti et al., 2010). In this method, proteins are first digested with soluble trypsin. Subsequently, proteolytic peptides are incubated with $H_2^{18}O$ at pH 4.5 in the presence of immobilized trypsin. Clearly, no singly [18]O labeled variants were observed in any of the peptide mass spectra (see Figure 5), indicating that no partial labeling whatsoever occurred, nor did any back exchange from [18]O to [16]O take place during sample treatment or analysis. Thus, complete incorporation of two [18]O labels into each of the tryptic peptides in a mixture can be achieved routinely.

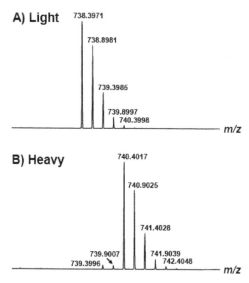

Fig. 5. Doubly charged tryptic peptide FLEQQNQVLQTK A) in the absence of [18]O label and B) after incorporation of the label. The two-step labeling reaction in the presence of immobilized trypsin as described here ultimately results in the complete incorporation of two [18]O labels, with no intermediary products present. The peptide isotope peaks in B) at $m/z$ 739.39 and 739.90 are due to impurities of commercial $H_2^{18}O$, containing only 97% [18]O.

It was shown in this study that [18]O labeling can be applied in a functional proteomics assay to discriminate background proteins from specific interactors of a protein of interest. Generally, controls are heavy labeled and the coimunoprecipitation (co-IP) sample is labeled light. Specific interactors are expected not to be present in the control and would thus have a ratio of (close to) zero, whereas background proteins would show heavy-to-light (H:L) ratios of (close to) 1. [18]O labeling was used in order to differentiate between non-specific background proteins and specific, *bona fide* interactors of the Cyclin dependent kinase 9

(Cdk9) purified from nuclear extracts of murine erythroleukemia (MEL) cells. Biotinylated Cdk9 was expressed in MEL cells and purified using streptavidin beads under relatively mild conditions (de Boer et al., 2003). The proteins that co-purified with Cdk9 were washed and digested with trypsin while still bound to the beads and subsequently identified by tandem mass spectrometry. A control sample was taken following the same procedure from an equal number of cells, but using non-transfected MEL cells. Proteolytic peptides from the control sample were then labeled using $H_2{}^{18}O$ in the two-step approach mentioned earlier, while proteolytic peptides from the Cdk9 pulldown sample underwent the same procedure with unlabeled $H_2O$. The peptide mixtures were dissolved in equal volumes of buffer, mixed in a 1:1 volume ratio and identified by LC-MS/MS. H:L ratios were calculated for all proteins identified from the mixed sample.

As expected, H:L ratios of close to 1 were observed for typical background proteins, such as ribosomal, housekeeping, and structural proteins, which were present as non-specific background proteins (see Figure 6). In contrast, among the proteins that were quantified with H:L ratios close to 0, indicating specificity for the Cdk9 co-immunopurification sample, the far majority of interacting proteins that have been described in different studies in the literature were identified in a single experiment, as well as several novel interaction partners of diverse functionalities, suggesting putative additional roles for Cdk9 in various nuclear events such as transcription and cell cycle control (Bezstarosti et al., 2010). It was shown in this study that complete $^{18}O$ labeling of peptides in complex mixtures can be routinely achieved. This greatly simplifies the analysis of peak intensity ratios, since only two components (*i.e.*, 'light' and 'heavy') need to be considered and no correction algorithms have to be applied to convert peak intensities of intermediately labeled peptide species.

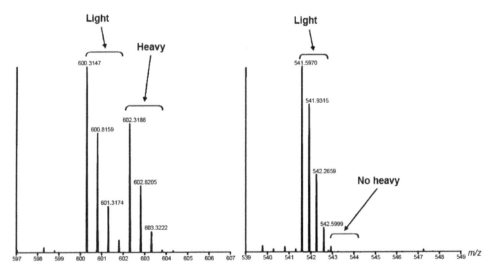

Fig. 6. MS spectra of two tryptic peptides from a 1:1 mixture of a digest of a Cdk9 co-IP experiment ($H_2{}^{16}O$) and a control sample (in $H_2{}^{18}O$). (A) Doubly charged peptide LGTPELSPTER of the contaminant acetyl-CoA carboxylase shows both the "light" and "heavy" forms and is therefore marked as a nonspecific protein. (B) Triply charged peptide GPPEETGAAVFDHPAK of cyclin T1 is only present in the "light" form and is therefore specific for the Cdk9 sample (see (Bezstarosti et al., 2010)).

## 2.2.4 Labeling with isobaric tags

Metabolic labeling, ICAT, enzymatic labeling and most other chemical labeling approaches for relative quantification are based on the mass difference between differentially labeled peptides. There are, however, some limitations imposed by mass difference labeling. The mass difference concept of many practical purposes is limited to a binary (2-plex) or ternary (3-plex) set of reagents; higher order multiplexing would increase the complexity of MS[1] spectra too much. This limitation makes comparison of multiple states difficult to undertake. Therefore, multiplexed sets of reagents for quantitative protein analysis have been developed. The isobaric tag for relative and absolute quantitation (iTRAQ) (Ross et al., 2004) and tandem mass tag (TMT) (Thompson et al., 2003) technologies are commercially available isobaric mass tagging reagents and protocols (Figure 7).

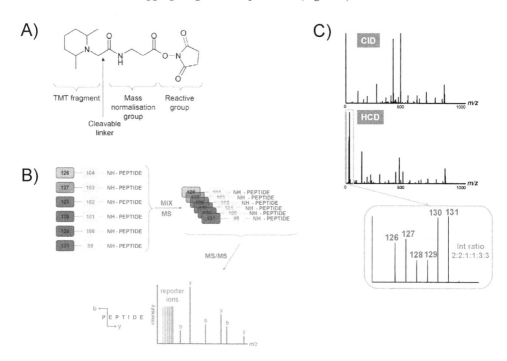

Fig. 7. A) Chemical structure of the TMT tag. The 6-plex tags have different distributions of the stable heavy isotopes of carbon and nitrogen in the molecule, resulting in different fragmentation spectra. B) A peptide that is present in 6 different samples is differentially labeled with a 6-plex TMT tag, containing reporter-balancer combinations, resulting in all conjugated peptides having the same $m/z$ value. Upon high energy collision dissociation (HCD), the differentially labeled peptides show identical b and y fragment ions, but the reporter ion masses in the low $m/z$ region are different. C) As an example, a protein was labeled in a 6-plex (TMT-126 through TMT-131) protocol and mixed in a 2:2:1:1:3:3 ratio. The resulting reporter ion intensity ratios show an excellent correlation with the mixing ratios. Panels A and B were adapted from (Thompson et al., 2003).

In these procedures, both N-termini and lysine side chains of peptides in a digest mixture are labeled with different isobaric mass reagents in such a way that all derivatized peptides are isobaric and chromatographically indistinguishable. Only upon peptide fragmentation can the different mass tags be distinguished. As each tag adds the same total mass to a given peptide, each peptide species produces only a single peak during liquid chromatography, even when two or more samples are mixed. Thus, there will be only one peak in the $MS^1$ scan, and, therefore, only a single $m/z$ will be isolated for fragmentation. The different mass tags only separate upon fragmentation, when reporter ions that are typical for each of the different labels are generated. These reporter ions are in the low mass range, which usually is not covered by typical peptide fragment ions. The intensity ratio of the different reporter ions is used as a quantitative readout. Thus, quantitation in combination with isobaric mass tagging is based on peptide fragmentation ($MS^2$) spectra rather than on the survey scans and quantitative accuracy will depend on the isolation width of precursor ions for fragmentation, since all ions isolated in that window will contribute to fragments in the reporter ion mass ranges. One drawback of such a method is that often only a single fragmentation spectrum per peptide is available, while in quantitation based $MS^1$ scans, usually several data points across the eluting peptide peak are sampled, which may result in a lower overall sensitivity.

### 2.3 Label free quantitation

Over the past few years, mainly as a result of constantly improving LC-MS equipment, there has been growing interest in the use of label-free approaches for quantitative proteomic analysis (see (Neilson et al., 2011) for a recent review). In a label free quantitative proteomic analysis, protein mixtures are analyzed directly and samples are compared to each other after independent analyses. As a result, there is no mixing of samples, so that higher proteome coverage can be achieved and there is no limit to the number of experiments that can be compared (Bantscheff et al., 2007). The disadvantage of this approach is a lack of a formal internal standard, which can lead to greater error in individual datasets but is minimized through the analysis of several biological replicates.

Label-free approaches may be divided into two main groups by the way that the abundance of a peptide is measured. The first group comprises methods that are based on the ion count and compare either maximum abundance or volume of ion count for peptide peaks at specific retention times between different samples (Chelius & Bondarenko, 2002; Listgarten & Emili, 2005; Silva et al., 2005; Wiener et al., 2004). As ionized peptides elute from a reversed-phase column into the mass spectrometer, their ion intensities can be measured within the given detection limits of the experimental setup. Although this method is relatively straightforward conceptually, several considerations must be taken into account to ensure reproducible and accurate detection and quantitation between individual sample runs. Concerns with LC signal resolution can arise when peptide signals are spread over a large retention time range causing overlap with co-eluting peptides. Similar concerns include biological variations resulting in multiple signals for the same peptide as well as technical variations in retention time, MS intensity, and sample background noise from chemical interference. These aspects of quantitation based on 'area under the curve' necessitate a computational 'clean up' of the raw LC-MS data (Neilson et al., 2011).

The second group is based on the identification of peptides by MS/MS and uses sampling statistics such as peptide count, spectral counts (Lundgren et al., 2010), or sequence coverage to quantify the differences between samples (Choi et al., 2008; Liu et al., 2004; Old et al.,

2005; Rappsilber et al., 2002). For protein quantification based on spectrum counting, the data processing steps are basically identical to the general protein identification workflow in proteomics, which is one of the reasons why this approach has become so popular. The rationale behind this quantitation method is that more abundant proteins are sampled more often in fragment ion scans than are low abundance peptides or proteins. Obviously, the outcome of spectrum counting depends on the settings of data-dependent acquisition on the mass spectrometer. In particular the linear range for quantitation and the number of proteins to be quantified are influenced by different settings for dynamic exclusion (Wang & Li, 2008); the optimal settings will depend on sample complexity. The most significant disadvantage of spectrum counting is that it behaves very poorly with proteins of low abundance and few spectra. The accuracy of the spectrum count method, especially for low abundance proteins, suffers from the fact that each spectrum is scored independently of its ion intensities.

## 3. Comparison of different methods for quantitation

With the existence of a wide variety of LC-MS based quantitation methods, it may be hard to decide which approach to utilize for a certain application. As described earlier, each approach has its own strengths and limitations, and, additionally, other factors may play a role, such as available equipment, level of experience and budget. In the following section we summarize the pros and cons of earlier described quantitation methods which might serve as a guidance to decide which approach is most suitable in a specific situation.

### 3.1 Metabolic labeling

If it is possible to label samples metabolically, this would be the most advantageous option to quantitate proteins. The most important reason for this is that different samples can be combined at the level of intact cells, which, as a result, excludes all sources of quantitation error introduced by biochemical and mass spectrometric procedures, as these will affect both protein populations in the same way. Metabolic labeling is therefore the most sensitive MS based labeling technique to date, making it possible to study small protein abundance differences as small as 1.5-fold changes or even smaller. Despite a number of cases that demonstrate the feasibility of metabolic labeling of higher organisms using $^{15}N$ sources *in vivo*, such as *C. elegans*, *D. Melanogaster* (Krijgsveld et al., 2003) and the rat (Wu et al., 2004), it is not practical to apply this strategy routinely. The most important reason for this is that labeling with $^{15}N$ complicates data analysis to a large extent, as discussed in section 2.1. Nowadays, the most widely applied method to metabolically label material of eukaryotic origin is SILAC in immortalized cell lines. SILAC based MS has been extensively applied for the study of global proteomes, in the field of functional proteomics, and for the analysis of post-translational modifications. Additionally, SILAC can be applied to whole organisms, such as *E. coli*, *S. cerevisiae*, and *D. melanogaster*. Even metabolic labeling of higher eukaryotes like the mouse (Kruger et al., 2008) has shown to be possible.

Although SILAC is the most accurate MS based quantitation approach, it might not always be possible or preferable to use SILAC. As mentioned earlier, not every cell type might grow well in the SILAC medium. Some cell lines readily convert arginine to proline, which complicates data analysis, and require adaptation of the protocol such as titration of arginine in the medium (Ong et al., 2003). Otherwise, computational approaches to correct arginine-to-proline conversion may be applied (Park et al., 2009). Finally, cell lines that are sensitive to changes in media composition or are otherwise difficult to grow or maintain in culture may not be amenable to metabolic labeling at all. When it is not possible to label a cell culture in SILAC medium, post-digestion incorporation methods may serve as an alternative. Moreover, post-digestion labeling might be the preferred method for affinity purification mass spectrometry (AP-MS) applications, as the starting material for co-IP assays is typically several milligrams of proteins. The use of stable isotope labeling by SILAC can be cost-prohibitive, whereas post-digestion labeling approaches such as stable isotope dimethyl labeling and $^{18}O$ labeling are performed with inexpensive generic reagents and do not pose severe financial restrictions to the amount of sample to be labeled.

In conclusion, SILAC can be applied in almost all sorts of proteomic applications since it is very sensitive, and limitations are mainly biological applicability or involve practical issues such as time, cost, or available equipment.

## 3.2 Chemical and enzymatic post-digestion labeling

One of the advantages of a chemical modification approach over metabolic labeling is the ability to label proteins after cell lysis and in a post-digestion manner. This makes the approach generically applicable, since it allows the quantitative analysis of biological samples that cannot be grown in culture, such as human body fluids or human tissue. ICAT was one of the first chemical labeling methods introduced for quantitative mass spectrometry. Although often and successfully applied, its main drawbacks are adverse side reactions and its inability to label peptides that do not contain cysteine residues. As a result, in many laboratories, ICAT has been substituted by other approaches, such as chemical dimethyl labeling or enzymatic $^{18}O$ labeling. Compared to ICAT, both $^{18}O$ labeling and dimethyl labeling are simple, free of extensive sample manipulations, virtually free of side reactions, and amenable to all protein species (*i.e.*, proteins that contain no cysteine residues). In contrast to ICAT, there is no lower limit of the protein amount that can be labeled for $^{18}O$ and dimethyl. Another advantage of the latter two labeling approaches is that they are cost-effective. This, together with the fact that proteins for any species can be labeled and the ease of sample preparation, makes chemical labeling the preferred method for the quantitative analysis of for instance size-limited human tissue specimens. Also, post-digestion labeling is practical for tissue samples of higher organisms such as mice, or cell lines that cannot be metabolically labeled.

One drawback of dimethyl labeling is that deuterated peptides show a small but significant retention time difference in reversed-phase HPLC compared to their non-deuterated counterparts (Zhang et al., 2001). This complicates data analysis because the relative quantities of the two peptide species cannot be determined accurately from one spectrum but requires integration across the chromatographic time scale. Retention time shifts are far less pronounced for labels such as $^{13}C$, $^{15}N$, or $^{18}O$ isotopes (Zhang & Regnier, 2002), so that the additional signal integration step over retention time can generally be omitted in approaches based on incorporation of these labels. However, compared to iTRAQ and TMT, dimethyl labeling is performed with inexpensive generic reagents and do not pose severe financial restrictions to the amount of sample to be labeled.

Multiplex labeling using TMT or iTRAQ has turned out to be particularly useful for following biological systems over multiple time points or, more generally, for comparing multiple treatments in the same experiment. With dimethylation labeling, iTRAQ and TMT labeling, multiplexing can be achieved, which is not possible for $^{18}O$ labeling. iTRAQ is capable of simultaneously analyzing eight samples (Pierce et al., 2008), whereas with TMT labeling, six samples can be measured together (Thompson et al., 2003). It should be noted that the use of commercial isobaric iTRAQ or TMT labels can be cost-prohibitive. In terms of equipment, TMT and iTRAQ labeling approaches are limited to mass spectrometers which are capable of efficiently detecting ions that are present at a relatively low $m/z$ and peptide quantification is based on a single fragmentation mass spectrum. An advantage of isobaric tags is that the labeled peptides co-elute from the chromatographic column which means that the MS signal is not split into different peaks, as in conventional isotope labeling, improving sensitivity in the MS mode.

In conclusion, chemical and enzymatic labeling can be applied to virtually any biological sample since incorporation is performed after cell lysis and generally also after digestion. Therefore, post-digestion labeling is specifically useful for the study of mammalian and human tissue or body fluids. Importantly, compared to metabolic labeling, label incorporation in chemical and enzymatic labeling approaches takes place at a later stage in the sample treatment protocol and are therefore in general less accurate. For absolute protein quantitation, peptides have to be labeled with stable heavy isotopes, which is usually done by synthesizing them with labeled amino acids, in order to serve as an internal standard.

### 3.3 Label free approaches

Since no labels are used whatsoever in label free quantitative proteomics, these approaches are inexpensive, they can be applied to any kind of biological material and the proteome coverage of quantified proteins is high because basically every protein that is identified by at least one peptide spectrum can in principle be quantified. In addition, the complexity of the sample is not increased by mixing different samples. Label free methods therefore usually have a high analytical depth and dynamic range, giving this method an advantage when large, global protein changes between treatments are expected. Also, since the samples are not mixed and quantification is done after MS analysis, the obtained data is not fixed and can be used in other contexts as well. These advantages make label free quantification an attractive approach for *e.g.* clinicians who have large patient material-derived datasets and want to compare multiple datasets, and have no wet lab available.

Despite the many advantages of label free quantitation, it is probably the least accurate among the mass spectrometric quantification methods when considering the overall experimental process because all the systematic and non-systematic variations between experiments are reflected in the obtained data. Consequently, the number of experimental steps should be kept to a minimum and every effort should be made to control reproducibility at each step.

There has been growing interest in the use of label-free approaches for quantitative proteomic analyses over the recent years, particularly because of ever increasing accuracy and reproducibility of high-resolution LC-MS equipment. Most MS analysis is performed with data dependent analysis (DDA) where the mass spectrometer runs a parent ion scan and selects the most abundant ions on which to conduct fragmentation scans, typically 4-10 scans, before returning to a parent ion scan. There may be a bias in this type of data for co-eluting peptides towards omitting the lower abundant peptides from MS/MS (Venable et al., 2004). This bias creates a subset of proteins effectively unseen due to the resultant level of detection limit. An experimental setup has been developed in which the mass spectrometer no longer cycles between MS and MS/MS mode but aims to detect and fragment all peptides in a chromatographic window simultaneously by rapidly alternating between high- and low-energy conditions in the mass spectrometer (Silva et al., 2006). Obviously, there are challenges with analyzing such data from complex samples as many fragmentation spectra will be populated with sequence ions from multiple peptides each contributing differently to the overall spectral content.

Also, there is evidence that label-free methods provide higher dynamic range of quantification than any stable isotope labeling approach (*i.e.*, 2-3 orders of magnitude) and therefore may be advantageous when large and global protein changes between experiments are observed (Old et al., 2005).

## 4. Data analysis

No matter the choice of quantitative method, quantitative proteomic data are typically very complex and often of variable quality. The main challenge stems from incomplete data, since even today's most advanced mass spectrometers cannot sample and fragment every peptide ion present in complex samples. As a consequence, only a subset of peptides and proteins present in a sample can be identified. Over the past years, a series of experimental strategies for mass spectrometry based quantitative proteomics and corresponding computational methodology for the processing of quantitative data have been generated (reviewed in (Matthiesen et al., 2011; Mueller et al., 2008). Conceptually different methods to perform quantitative LC-MS experiments demand different quantification principles and available software solutions for data analysis. Quantification can be achieved by comparing peak intensities in differential stable isotopic labeling, via spectral counting, or by using the ion current in label-free LC-MS measurements. Numerous software solutions have been presented, with specific instrument compatibility and processing functionality and which can cope with these basically different quantitation methods. It is important for researchers to choose an appropriate software solution for quantitative proteomic experiments based on their experimental and analytical requirements. However, it goes beyond the scope of this Chapter to discuss all of the available software tools separately. For an extensive and up-to-date overview of software solutions including links to websites for downloads, the reader is referred to http://www.ms-utils.org.

## 5. Concluding remarks

As we have discussed in this Chapter, all of the mass spectrometry based quantification methods have their particular strengths and weaknesses. The researcher has to choose the best method from the multitude of methods that have emerged for the analysis of simple and complex (sub-)proteomes using quantitative mass spectrometry for his or her specific research; a choice that depends on the financial aspects involved, the availability of high-resolution mass spectrometer and LC equipment and the available expertise present in the lab. Quantitative proteomics methods are now starting to mature to an extent that they can be meaningfully applied to the study of proteomes and their dynamics. Using the labeling methods described in this Chapter, it is now possible to identify and quantitate several thousands of proteins in a single experiment. However, there is still room for significant improvements to the experimental strategies that are required for the quantitative analysis of very complex mixtures and of post-translational modifications, with the ultimate aim to generate quantitative proteomic data at a scale which would allow the comprehensive investigation of a biological phenomenon. At the same time, the recent exponential increase in data volume and complexity demands the development of appropriate bioinformatic and statistical approaches in order to arrive at meaningful interpretations of the results. This can only be achieved if the influence of the employed technologies on the results obtained is well understood.

## 6. References

Bantscheff, M., Schirle, M., Sweetman, G., Rick, J. & Kuster, B. (2007). Quantitative Mass Spectrometry in Proteomics: A Critical Review. Anal Bioanal Chem, 389(4), 1017-1031.

Bezstarosti, K., Ghamari, A., Grosveld, F. G. & Demmers, J. A. (2010). Differential Proteomics Based on 18o Labeling to Determine the Cyclin Dependent Kinase 9 Interactome. J Proteome Res, 9(9), 4464-4475.

Blagoev, B., Kratchmarova, I., Ong, S. E., Nielsen, M., Foster, L. J. & Mann, M. (2003). A Proteomics Strategy to Elucidate Functional Protein-Protein Interactions Applied to Egf Signaling. Nat Biotechnol, 21(3), 315-318.

Blagoev, B. & Mann, M. (2006). Quantitative Proteomics to Study Mitogen-Activated Protein Kinases. Methods, 40(3), 243-250.

Blonder, J., Hale, M. L., Chan, K. C., Yu, L. R., Lucas, D. A., Conrads, T. P., Zhou, M., Popoff, M. R., Issaq, H. J., Stiles, B. G. & Veenstra, T. D. (2005). Quantitative Profiling of the Detergent-Resistant Membrane Proteome of Iota-B Toxin Induced Vero Cells. J Proteome Res, 4(2), 523-531.

Boersema, P. J., Aye, T. T., van Veen, T. A., Heck, A. J. & Mohammed, S. (2008). Triplex Protein Quantification Based on Stable Isotope Labeling by Peptide Dimethylation Applied to Cell and Tissue Lysates. Proteomics, 8(22), 4624-4632.

Boersema, P. J., Foong, L. Y., Ding, V. M., Lemeer, S., van Breukelen, B., Philp, R., Boekhorst, J., Snel, B., den Hertog, J., Choo, A. B. & Heck, A. J. (2010). In-Depth Qualitative and Quantitative Profiling of Tyrosine Phosphorylation Using a Combination of Phosphopeptide Immunoaffinity Purification and Stable Isotope Dimethyl Labeling. Mol Cell Proteomics, 9(1), 84-99.

Boersema, P. J., Raijmakers, R., Lemeer, S., Mohammed, S. & Heck, A. J. (2009). Multiplex Peptide Stable Isotope Dimethyl Labeling for Quantitative Proteomics. Nat Protoc, 4(4), 484-494.

Bonenfant, D., Schmelzle, T., Jacinto, E., Crespo, J. L., Mini, T., Hall, M. N. & Jenoe, P. (2003). Quantitation of Changes in Protein Phosphorylation: A Simple Method Based on Stable Isotope Labeling and Mass Spectrometry. Proc Natl Acad Sci U S A, 100(3), 880-885.

Brown, K. J. & Fenselau, C. (2004). Investigation of Doxorubicin Resistance in Mcf-7 Breast Cancer Cells Using Shot-Gun Comparative Proteomics with Proteolytic 18o Labeling. J Proteome Res, 3(3), 455-462.

Chelius, D. & Bondarenko, P. V. (2002). Quantitative Profiling of Proteins in Complex Mixtures Using Liquid Chromatography and Mass Spectrometry. J Proteome Res, 1(4), 317-323.

Chen, X., Cushman, S. W., Pannell, L. K. & Hess, S. (2005). Quantitative Proteomic Analysis of the Secretory Proteins from Rat Adipose Cells Using a 2d Liquid Chromatography-Ms/Ms Approach. J Proteome Res, 4(2), 570-577.

Choi, H., Fermin, D. & Nesvizhskii, A. I. (2008). Significance Analysis of Spectral Count Data in Label-Free Shotgun Proteomics. Mol Cell Proteomics, 7(12), 2373-2385.

Conrads, T. P., Alving, K., Veenstra, T. D., Belov, M. E., Anderson, G. A., Anderson, D. J., Lipton, M. S., Pasa-Tolic, L., Udseth, H. R., Chrisler, W. B., Thrall, B. D. & Smith, R. D. (2001). Quantitative Analysis of Bacterial and Mammalian Proteomes Using a Combination of Cysteine Affinity Tags and 15n-Metabolic Labeling. Anal Chem, 73(9), 2132-2139.

de Boer, E., Rodriguez, P., Bonte, E., Krijgsveld, J., Katsantoni, E., Heck, A., Grosveld, F. & Strouboulis, J. (2003). Efficient Biotinylation and Single-Step Purification of Tagged Transcription Factors in Mammalian Cells and Transgenic Mice. Proc Natl Acad Sci U S A, 100(13), 7480-7485.

de Godoy, L. M., Olsen, J. V., de Souza, G. A., Li, G., Mortensen, P. & Mann, M. (2006). Status of Complete Proteome Analysis by Mass Spectrometry: Silac Labeled Yeast as a Model System. Genome Biol, 7(6), R50.

Engelsberger, W. R., Erban, A., Kopka, J. & Schulze, W. X. (2006). Metabolic Labeling of Plant Cell Cultures with K(15)No3 as a Tool for Quantitative Analysis of Proteins and Metabolites. Plant Methods, 2, 14.

Fenselau, C. & Yao, X. (2007). Proteolytic Labeling with 18o for Comparative Proteomics Studies: Preparation of 18o-Labeled Peptides and the 18o/16o Peptide Mixture. Methods Mol Biol, 359, 135-142.

Gevaert, K., Staes, A., Van Damme, J., De Groot, S., Hugelier, K., Demol, H., Martens, L., Goethals, M. & Vandekerckhove, J. (2005). Global Phosphoproteome Analysis on Human Hepg2 Hepatocytes Using Reversed-Phase Diagonal Lc. Proteomics, 5(14), 3589-3599.

Gruhler, A., Schulze, W. X., Matthiesen, R., Mann, M. & Jensen, O. N. (2005). Stable Isotope Labeling of Arabidopsis Thaliana Cells and Quantitative Proteomics by Mass Spectrometry. Mol Cell Proteomics, 4(11), 1697-1709.

Gygi, S. P., Rist, B., Gerber, S. A., Turecek, F., Gelb, M. H. & Aebersold, R. (1999). Quantitative Analysis of Complex Protein Mixtures Using Isotope-Coded Affinity Tags. Nat Biotechnol, 17(10), 994-999.

Hajkova, D., Rao, K. C. & Miyagi, M. (2006). Ph Dependency of the Carboxyl Oxygen Exchange Reaction Catalyzed by Lysyl Endopeptidase and Trypsin. J Proteome Res, 5(7), 1667-1673.

Hansen, K. C., Schmitt-Ulms, G., Chalkley, R. J., Hirsch, J., Baldwin, M. A. & Burlingame, A. L. (2003). Mass Spectrometric Analysis of Protein Mixtures at Low Levels Using Cleavable 13c-Isotope-Coded Affinity Tag and Multidimensional Chromatography. Mol Cell Proteomics, 2(5), 299-314.

Hood, B. L., Darfler, M. M., Guiel, T. G., Furusato, B., Lucas, D. A., Ringeisen, B. R., Sesterhenn, I. A., Conrads, T. P., Veenstra, T. D. & Krizman, D. B. (2005). Proteomic Analysis of Formalin-Fixed Prostate Cancer Tissue. Mol Cell Proteomics, 4(11), 1741-1753.

Hood, B. L., Lucas, D. A., Kim, G., Chan, K. C., Blonder, J., Issaq, H. J., Veenstra, T. D., Conrads, T. P., Pollet, I. & Karsan, A. (2005). Quantitative Analysis of the Low

Molecular Weight Serum Proteome Using 18o Stable Isotope Labeling in a Lung Tumor Xenograft Mouse Model. J Am Soc Mass Spectrom, 16(8), 1221-1230.

Hsu, J. L., Huang, S. Y., Chow, N. H. & Chen, S. H. (2003). Stable-Isotope Dimethyl Labeling for Quantitative Proteomics. Anal Chem, 75(24), 6843-6852.

Ibarrola, N., Kalume, D. E., Gronborg, M., Iwahori, A. & Pandey, A. (2003). A Proteomic Approach for Quantitation of Phosphorylation Using Stable Isotope Labeling in Cell Culture. Anal Chem, 75(22), 6043-6049.

Ippel, J. H., Pouvreau, L., Kroef, T., Gruppen, H., Versteeg, G., van den Putten, P., Struik, P. C. & van Mierlo, C. P. (2004). In Vivo Uniform (15)N-Isotope Labelling of Plants: Using the Greenhouse for Structural Proteomics. Proteomics, 4(1), 226-234.

Krijgsveld, J., Ketting, R. F., Mahmoudi, T., Johansen, J., Artal-Sanz, M., Verrijzer, C. P., Plasterk, R. H. & Heck, A. J. (2003). Metabolic Labeling of C. Elegans and D. Melanogaster for Quantitative Proteomics. Nat Biotechnol, 21(8), 927-931.

Kruger, M., Moser, M., Ussar, S., Thievessen, I., Luber, C. A., Forner, F., Schmidt, S., Zanivan, S., Fassler, R. & Mann, M. (2008). Silac Mouse for Quantitative Proteomics Uncovers Kindlin-3 as an Essential Factor for Red Blood Cell Function. Cell, 134(2), 353-364.

Lanquar, V., Kuhn, L., Lelievre, F., Khafif, M., Espagne, C., Bruley, C., Barbier-Brygoo, H., Garin, J. & Thomine, S. (2007). 15n-Metabolic Labeling for Comparative Plasma Membrane Proteomics in Arabidopsis Cells. Proteomics, 7(5), 750-754.

Li, J., Steen, H. & Gygi, S. P. (2003). Protein Profiling with Cleavable Isotope-Coded Affinity Tag (Cicat) Reagents: The Yeast Salinity Stress Response. Mol Cell Proteomics, 2(11), 1198-1204.

Listgarten, J. & Emili, A. (2005). Statistical and Computational Methods for Comparative Proteomic Profiling Using Liquid Chromatography-Tandem Mass Spectrometry. Mol Cell Proteomics, 4(4), 419-434.

Liu, H., Sadygov, R. G. & Yates, J. R., 3rd. (2004). A Model for Random Sampling and Estimation of Relative Protein Abundance in Shotgun Proteomics. Anal Chem, 76(14), 4193-4201.

Lundgren, D. H., Hwang, S. I., Wu, L. & Han, D. K. (2010). Role of Spectral Counting in Quantitative Proteomics. Expert Rev Proteomics, 7(1), 39-53.

Matthiesen, R., Azevedo, L., Amorim, A. & Carvalho, A. S. (2011). Discussion on Common Data Analysis Strategies Used in Ms-Based Proteomics. Proteomics, 11(4), 604-619.

McClatchy, D. B., Dong, M. Q., Wu, C. C., Venable, J. D. & Yates, J. R., 3rd. (2007). 15n Metabolic Labeling of Mammalian Tissue with Slow Protein Turnover. J Proteome Res, 6(5), 2005-2010.

Miyagi, M. & Rao, K. C. (2007). Proteolytic 18o-Labeling Strategies for Quantitative Proteomics. Mass Spectrom Rev, 26(1), 121-136.

Mortensen, P., Gouw, J. W., Olsen, J. V., Ong, S. E., Rigbolt, K. T., Bunkenborg, J., Cox, J., Foster, L. J., Heck, A. J., Blagoev, B., Andersen, J. S. & Mann, M. (2010). Msquant, an Open Source Platform for Mass Spectrometry-Based Quantitative Proteomics. J Proteome Res, 9(1), 393-403.

Mueller, L. N., Brusniak, M. Y., Mani, D. R. & Aebersold, R. (2008). An Assessment of Software Solutions for the Analysis of Mass Spectrometry Based Quantitative Proteomics Data. J Proteome Res, 7(1), 51-61.

Neilson, K. A., Ali, N. A., Muralidharan, S., Mirzaei, M., Mariani, M., Assadourian, G., Lee, A., van Sluyter, S. C. & Haynes, P. A. (2011). Less Label, More Free: Approaches in Label-Free Quantitative Mass Spectrometry. Proteomics, 11(4), 535-553.

Oda, Y., Huang, K., Cross, F. R., Cowburn, D. & Chait, B. T. (1999). Accurate Quantitation of Protein Expression and Site-Specific Phosphorylation. Proc Natl Acad Sci U S A, 96(12), 6591-6596.

Old, W. M., Meyer-Arendt, K., Aveline-Wolf, L., Pierce, K. G., Mendoza, A., Sevinsky, J. R., Resing, K. A. & Ahn, N. G. (2005). Comparison of Label-Free Methods for Quantifying Human Proteins by Shotgun Proteomics. Mol Cell Proteomics, 4(10), 1487-1502.

Ong, S. E., Blagoev, B., Kratchmarova, I., Kristensen, D. B., Steen, H., Pandey, A. & Mann, M. (2002). Stable Isotope Labeling by Amino Acids in Cell Culture, Silac, as a Simple and Accurate Approach to Expression Proteomics. Mol Cell Proteomics, 1(5), 376-386.

Ong, S. E., Kratchmarova, I. & Mann, M. (2003). Properties of 13c-Substituted Arginine in Stable Isotope Labeling by Amino Acids in Cell Culture (Silac). J Proteome Res, 2(2), 173-181.

Ong, S. E. & Mann, M. (2005). Mass Spectrometry-Based Proteomics Turns Quantitative. Nat Chem Biol, 1(5), 252-262.

Park, S. K., Liao, L., Kim, J. Y. & Yates, J. R., 3rd. (2009). A Computational Approach to Correct Arginine-to-Proline Conversion in Quantitative Proteomics. Nat Methods, 6(3), 184-185.

Pierce, A., Unwin, R. D., Evans, C. A., Griffiths, S., Carney, L., Zhang, L., Jaworska, E., Lee, C. F., Blinco, D., Okoniewski, M. J., Miller, C. J., Bitton, D. A., Spooncer, E. & Whetton, A. D. (2008). Eight-Channel Itraq Enables Comparison of the Activity of Six Leukemogenic Tyrosine Kinases. Mol Cell Proteomics, 7(5), 853-863.

Qian, W. J., Monroe, M. E., Liu, T., Jacobs, J. M., Anderson, G. A., Shen, Y., Moore, R. J., Anderson, D. J., Zhang, R., Calvano, S. E., Lowry, S. F., Xiao, W., Moldawer, L. L., Davis, R. W., Tompkins, R. G., Camp, D. G., 2nd & Smith, R. D. (2005). Quantitative Proteome Analysis of Human Plasma Following in Vivo Lipopolysaccharide Administration Using 16o/18o Labeling and the Accurate Mass and Time Tag Approach. Mol Cell Proteomics, 4(5), 700-709.

Raijmakers, R., Berkers, C. R., de Jong, A., Ovaa, H., Heck, A. J. & Mohammed, S. (2008). Automated Online Sequential Isotope Labeling for Protein Quantitation Applied to Proteasome Tissue-Specific Diversity. Mol Cell Proteomics, 7(9), 1755-1762.

Ramos-Fernandez, A., Lopez-Ferrer, D. & Vazquez, J. (2007). Improved Method for Differential Expression Proteomics Using Trypsin-Catalyzed 18o Labeling with a Correction for Labeling Efficiency. Mol Cell Proteomics, 6(7), 1274-1286.

Rao, K. C., Palamalai, V., Dunlevy, J. R. & Miyagi, M. (2005). Peptidyl-Lys Metalloendopeptidase-Catalyzed 18o Labeling for Comparative Proteomics: Application to Cytokine/Lipolysaccharide-Treated Human Retinal Pigment Epithelium Cell Line. Mol Cell Proteomics, 4(10), 1550-1557.

Rappsilber, J., Ryder, U., Lamond, A. I. & Mann, M. (2002). Large-Scale Proteomic Analysis of the Human Spliceosome. Genome Res, 12(8), 1231-1245.

Ross, P. L., Huang, Y. N., Marchese, J. N., Williamson, B., Parker, K., Hattan, S., Khainovski, N., Pillai, S., Dey, S., Daniels, S., Purkayastha, S., Juhasz, P., Martin, S., Bartlet-

Jones, M., He, F., Jacobson, A. & Pappin, D. J. (2004). Multiplexed Protein Quantitation in Saccharomyces Cerevisiae Using Amine-Reactive Isobaric Tagging Reagents. Mol Cell Proteomics, 3(12), 1154-1169.

Sevinsky, J. R., Brown, K. J., Cargile, B. J., Bundy, J. L. & Stephenson, J. L., Jr. (2007). Minimizing Back Exchange in 18o/16o Quantitative Proteomics Experiments by Incorporation of Immobilized Trypsin into the Initial Digestion Step. Anal Chem, 79(5), 2158-2162.

Silva, J. C., Denny, R., Dorschel, C., Gorenstein, M. V., Li, G. Z., Richardson, K., Wall, D. & Geromanos, S. J. (2006). Simultaneous Qualitative and Quantitative Analysis of the Escherichia Coli Proteome: A Sweet Tale. Mol Cell Proteomics, 5(4), 589-607.

Silva, J. C., Denny, R., Dorschel, C. A., Gorenstein, M., Kass, I. J., Li, G. Z., McKenna, T., Nold, M. J., Richardson, K., Young, P. & Geromanos, S. (2005). Quantitative Proteomic Analysis by Accurate Mass Retention Time Pairs. Anal Chem, 77(7), 2187-2200.

Staes, A., Demol, H., Van Damme, J., Martens, L., Vandekerckhove, J. & Gevaert, K. (2004). Global Differential Non-Gel Proteomics by Quantitative and Stable Labeling of Tryptic Peptides with Oxygen-18. J Proteome Res, 3(4), 786-791.

Sury, M. D., Chen, J. X. & Selbach, M. (2010). The Silac Fly Allows for Accurate Protein Quantification in Vivo. Mol Cell Proteomics, 9(10), 2173-2183.

Synowsky, S. A., van Wijk, M., Raijmakers, R. & Heck, A. J. (2009). Comparative Multiplexed Mass Spectrometric Analyses of Endogenously Expressed Yeast Nuclear and Cytoplasmic Exosomes. J Mol Biol, 385(4), 1300-1313.

Thompson, A., Schafer, J., Kuhn, K., Kienle, S., Schwarz, J., Schmidt, G., Neumann, T., Johnstone, R., Mohammed, A. K. & Hamon, C. (2003). Tandem Mass Tags: A Novel Quantification Strategy for Comparative Analysis of Complex Protein Mixtures by Ms/Ms. Anal Chem, 75(8), 1895-1904.

Unlu, M., Morgan, M. E. & Minden, J. S. (1997). Difference Gel Electrophoresis: A Single Gel Method for Detecting Changes in Protein Extracts. Electrophoresis, 18(11), 2071-2077.

Van Hoof, D., Pinkse, M. W., Oostwaard, D. W., Mummery, C. L., Heck, A. J. & Krijgsveld, J. (2007). An Experimental Correction for Arginine-to-Proline Conversion Artifacts in Silac-Based Quantitative Proteomics. Nat Methods, 4(9), 677-678.

Venable, J. D., Dong, M. Q., Wohlschlegel, J., Dillin, A. & Yates, J. R. (2004). Automated Approach for Quantitative Analysis of Complex Peptide Mixtures from Tandem Mass Spectra. Nat Methods, 1(1), 39-45.

Wang, N. & Li, L. (2008). Exploring the Precursor Ion Exclusion Feature of Liquid Chromatography-Electrospray Ionization Quadrupole Time-of-Flight Mass Spectrometry for Improving Protein Identification in Shotgun Proteome Analysis. Anal Chem, 80(12), 4696-4710.

Wiener, M. C., Sachs, J. R., Deyanova, E. G. & Yates, N. A. (2004). Differential Mass Spectrometry: A Label-Free Lc-Ms Method for Finding Significant Differences in Complex Peptide and Protein Mixtures. Anal Chem, 76(20), 6085-6096.

Wu, C. C., MacCoss, M. J., Howell, K. E., Matthews, D. E. & Yates, J. R., 3rd. (2004). Metabolic Labeling of Mammalian Organisms with Stable Isotopes for Quantitative Proteomic Analysis. Anal Chem, 76(17), 4951-4959.

Yao, X., Freas, A., Ramirez, J., Demirev, P. A. & Fenselau, C. (2001). Proteolytic 18o Labeling for Comparative Proteomics: Model Studies with Two Serotypes of Adenovirus. Anal Chem, 73(13), 2836-2842.

Zang, L., Palmer Toy, D., Hancock, W. S., Sgroi, D. C. & Karger, B. L. (2004). Proteomic Analysis of Ductal Carcinoma of the Breast Using Laser Capture Microdissection, Lc-Ms, and 16o/18o Isotopic Labeling. J Proteome Res, 3(3), 604-612.

Zhang, R. & Regnier, F. E. (2002). Minimizing Resolution of Isotopically Coded Peptides in Comparative Proteomics. J Proteome Res, 1(2), 139-147.

Zhang, R., Sioma, C. S., Wang, S. & Regnier, F. E. (2001). Fractionation of Isotopically Labeled Peptides in Quantitative Proteomics. Anal Chem, 73(21), 5142-5149.

# Part 3

# 2D Gel Electrophoresis and Databases

# 7

# 2D-PAGE Database for Studies on Energetic Metabolism of the Denitrifying Bacterium *Paracoccus denitrificans*

Pavel Bouchal[1,2], Robert Stein[3], Zbyněk Zdráhal[4],
Peter R. Jungblut[5] and Igor Kučera[1]
*[1]Department of Biochemistry, Faculty of Science,
Masaryk University, Brno
[2]Regional Centre for Applied Molecular Oncology,
Masaryk Memorial Cancer Institute, Brno
[3]I&B Informatics and Biology, Berlin
[4]Core Facility – Proteomics, Central European
Institute of Technology, Masaryk University, Brno
[5]Max Planck Institute for Infection Biology,
Core Facility Protein Analysis, Berlin
[1,2,4]Czech Republic
[3,5]Germany*

## 1. Introduction

The gram-negative soil bacterium *Paracoccus denitrificans* is a chemoorganotroph and a facultative chemolithotroph, capable of using the oxidation of molecular hydrogen, methanol or thiosulphate as sole source of energy for autotrophic growth. Many different organic compounds serve as sole carbon source, the metabolism is, however, always respiratory and never fermentative. *P. denitrificans* synthesizes three distinct terminal oxidases ($aa_3$-type and $cbb_3$-type cytochrome $c$ oxidases and $ba_3$-type quinol oxidase) during aerobic growth (Fig. 1). Under limited oxygen concentration, it can produce four additional terminal oxidoreductases for stepwise anaerobic conversion of nitrate to nitrogen gas (denitrification): nitrate reductase, nitrite reductase, nitrous oxide reductase and nitric oxide reductase (Fig. 2). Synthesis of these enzymes is tightly controlled at the transcription level: (i) globally according to an energetic hierarchy and (ii) on the level of the individual genes. As a result, a proper balance in the concentration and activity of these reductases is achieved and the cytotoxicity of the toxic intermediates of denitrification, nitrite and nitric oxide, is eliminated (Zumft 1997). The major players in the mentioned regulatory network are three members of the FNR (fumarate and nitrate reductase regulatory) protein family of transcription regulators. Upon activation by their corresponding signals, they bind to specific sites (FNR boxes) in target promoters upstream/downstream of the σ factor binding site and destabilize/stabilize the RNA-polymerase transcription initiation complex. The first regulatory protein is FnrP which has a [4Fe-4S] cluster for oxygen sensing, the second is

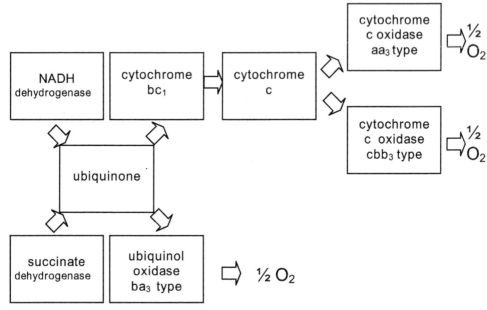

Fig. 1. Major components of *P. denitrificans* respiratory chain. Three distinct terminal oxidases are synthesized depending on environmental factors (*e.g.*, oxygen tension). The arrows indicate the electron flow. Inhibition of terminal oxidases with azide led to elevation of superoxide dismutase as revealed using a proteomic approach and confirmed at transcript and enzyme activity level (Bouchal *et al.* 2011).

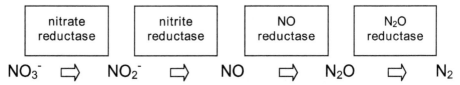

Fig. 2. Schema of *P. denitrificans* anaerobic denitrification pathway. Nitrate reductase β-subunit, nitrite reductase and nitrous oxide reductase can be detected and quantified using a proteomic approach (Bouchal *et al.* 2004; Bouchal *et al.* 2011). The expression of the denitrification enzymes is tightly controlled by FnrP, NNR and NarR transcription regulators at transcription level.

NNR, which has a heme for NO sensing and the third one is NarR which is poorly characterized and likely to be a nitrite sensor (Van Spanning *et al.* 1997; Wood *et al.* 2001). In response to oxygen deprivation, FnrP controls expression of the *nar* gene cluster encoding nitrate reductase, the *cco*-gene cluster encoding a *cbb₃*-type oxidase for respiration at low oxygen concentrations and the *ccp* gene encoding cytochrome *c* peroxidase. NNR specifically controls expression of the gene clusters encoding the nitrite (*nirS*), and nitric oxide (*norCB*) reductases and, to a certain extent, nitrous oxide (*nosZ*) reductase. NarR is required for transcription of the *nar* gene cluster in an unknown interplay with the FnrP protein (Wood *et al.* 2001; Veldman *et al.* 2006). These properties have been deduced from a number of studies on each of these transcriptional activators, but knowledge on the interplay between these regulators along with their position in the complete regulatory network are scarce.

Given its metabolic versatility, this bacterium becomes an excellent model system to study the mechanisms of cellular responses to different environments.

## 2. *P. denitrificans* proteome analysis: Method development

Our group gained extensive experience in studying *P. denitrificans* physiology, reaching from measurement of enzyme activities to characterization of the role of individual proteins, *e.g.*, pseudoazurin (Koutny & Kucera 1999), nitrate transporter (Kucera 2003) and ferric iron reductases (Mazoch *et al.* 2004; Sedlacek *et al.* 2009). New proteomic technologies have been employed in our research since 2001 when we used a two-dimensional gel electrophoresis (2D-PAGE) with carrier ampholytes for the first time. Using this technique, we obtained good separations of about 150 proteins present in membrane fraction, allowing a comparison of its protein composition with the periplasmic fraction (Bouchal & Kucera 2004), see also Fig. 3. However, the capacity of original sample preparation procedure (Bouchal & Kucera 2004) led to difficulties with identification of less intensive protein spots.

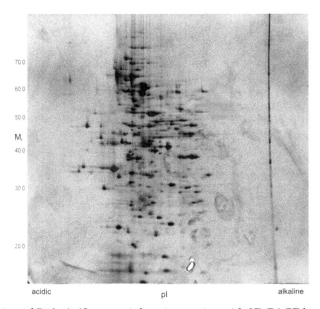

Fig. 3. A separation of *P. denitrificans* periplasmic proteins with 2D-PAGE based on carrier ampholytes–isoelectric focusing in the first dimension (Bouchal & Kucera 2002), with permission.

After establishing the mass spectrometry laboratory in 2002, we optimized new methods for the proteome analysis of *P. denitrificans* using immobilized pH gradients. See next paragraphs for complete protocols and Fig. 4 for a typical 2-D proteome map. Namely, the mass spectrometric analysis opened the way towards high-throughput and precise protein identification and valid conclusions made based on proteomics data. Matrix-Assisted Laser

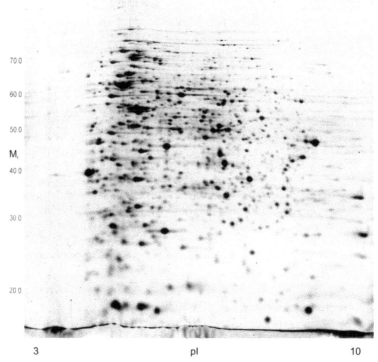

Fig. 4. Current stage of *P. denitrificans* proteomics approach: 2D-PAGE map (20 cm x 20 cm size) prepared with non-linear immobilized pH gradients (pH 3-10 NL) and Sypro Ruby staining has been used for both quantitative analysis and mass spectrometry protein identification.

Desorption-Ionization Mass Spectrometry (MALDI-MS) was used in initial proteomic studies identifying proteins exclusively by peptide mass fingerprinting. In addition, sensitivity of our MALDI-MS instrumentation of that time was not sufficient in case of weak protein spots. Since 2007, we have started to identify proteins using tandem mass spectrometric techniques, MALDI-MS/MS and ESI-MS/MS (concretely, capillary liquid chromatography – ion trap mass spectrometry with electrospray ionization), which resulted in more reliable protein identification based on MS/MS data. To improve the sensitivity of our LC-MS/MS system, we introduced nano-scale LC separation in 2008. At present, practically each analysis of protein spots leads to positive identification, involving sensitive fluorescent staining (Sypro Ruby). Subsequently, several comprehensive proteomic experiments were performed during the years using our proteomic platform in order to study the differences in protein composition caused by the growth on different terminal electron acceptors in both total cell lysates and membrane fractions (Bouchal *et al.* 2004). An additional large proteomic study with data confirmation at transcript level was performed to describe the regulons of three FNR-type transcription regulators FnrP, NNR and NarR at protein level (Bouchal *et al.* 2010). Quantitative and statistical image analysis primarily resulted in creation of local database files in a PDQUEST format. Subsequently, we decided to publish the 2-D maps in a web form to make all details accessible to other researchers online *via* the "Proteome Database System for Microbial Research at Max Planck Institute for Infection Biology" in Berlin, Germany as described below.

## 3. Optimized methods used for *P. denitrificans* proteomics

### 3.1 Bacteria and culture conditions

Four strains of *P. denitrificans* were used in the published studies: Pd1222 (wild type), Pd2921 (FnrP mutant (Van Spanning *et al.* 1997)), Pd7721 (NNR mutant (Van Spanning *et al.* 1995)) and Pd11021 (NarR mutant, unpublished data). These four strains were cultivated at 30 °C in 1 l bottles filled with 0.5 l cultures with a starting optical density at 600 nm of 0.01, under the three following growth conditions: (i) aerobically at 250 rpm up to an optical density of 0.6, (ii) semiaerobically at 100 rpm up to an optical density of 1.0 and (iii) semiaerobically with nitrate at 100 rpm up to an optical density of 1.0. The minimal medium was composed of $NH_4Cl$ (30 mM), sodium succinate (25 mM), $Na_2MoO_4$ (0.6 mM), $MgSO_4$ (0.4 mM), EDTA (0.25 mM), Lawford trace solution (1 mL/L) and potassium phosphate (65 mM, pH 7.0); $KNO_3$ (100 mM) was added in the case of cultivations in the presence of nitrate. Each culture was grown in three biological replicates, and as such we availed of a set of 36 independently grown *P. denitrificans* cultures. Cells were harvested by centrifugation (6 200 x g, 30 min), washed with 50 mM tris(hydroxymethyl)aminomethane/HCl (Tris/HCl) pH 7.3 and stored as a pellet at -80 °C.

### 3.2 Sample preparation and two-dimensional gel electrophoresis

After cultivation, the cells were disrupted by sonicating 15 mg (wet weight) of pellet for 30 x 0.1 s (50 W output) in 300 µL of lysis buffer containing 7 M urea, 2 M thiourea, 1% (w/v) (3-((4-Heptyl)phenyl-3-hydroxypropyl)dimethylammoniopropanesulfonate) (C7BzO), 40 mM Tris-base, 70 mM dithiothreitol (DTT), 2% (v/v) Pharmalyte 3/10, 5 mM NaF, 0.2 mM $NaVO_3$, CompleteMini Protease Inhibitor Cocktail (Roche, Penzberg, Germany, one tablet per 10 mL of lysis buffer) and 150 U of benzonase (Sigma-Aldrich, St. Louis, MO, USA). The cell extracts were incubated for 1.5 h at 20 °C. Cellular debris was then removed by centrifugation (16000 x g, 20 min, 15 °C) and the supernatant (total cell lysate) was stored at –80 °C.

For preparation of membrane fraction, cells were disintegrated and converted into membrane vesicles as previously described (Burnell *et al.* 1975) with several modifications. Briefly, the suspension of harvested cells was diluted with 5.7 volumes of a solution containing 0.5 M sucrose, 200 mM Tris/HCl pH 7.3, 0.5 mM EDTA and lysozyme (1 mg/ml of the total volume). After 45 min of enzymatic lysis at 30 °C, an osmotic lysis (45 min, 30 °C) was initiated by addition of equal volume of ice-chilled water. After centrifugation (4600 x g / 30 min/4 °C), the pellet (spheroplasts) was further lyzed (30 min/4 °C) in 7.5 volumes of chilled water containing a trace of DNAse and 14 mM $MgSO_4$. The unbroken spheroplasts were sedimented at 4600 x g (30 min/4 °C) and the supernatant from this step was subjected to ultracentrifugation at 184000 x g for 40 min (4°C) using Beckman L8-55M Ultracentrifuge with 45 Ti rotor (Beckman, USA). The collected membranes were resuspended in 50 mM Tris/HCl pH 7.3, utracentrifuged again and resuspended in the same buffer. 150 µg of protein for analytical gels or 1 mg of protein for micropreparative separations, respectively, were extracted using sample solution containing 7 M urea, 2 M thiourea, 1 % (w/v) 3-[N,N-dimethyl(3-myristoylaminopropyl)ammonio]propanesulfonate (ASB 14), 1 % (v/v) TRITON X-100, 2 mM tributylphosphine, 15 mM Tris base, 1 % (v/v) Pharmalyte 3/10 and 0.5 % (v/v) Pharmalyte 8/10 for 1.5 h at 20 °C.

The protein content was determined by RC-DC Protein Assay (Bio-Rad, Hercules, CA) with BSA as a standard. Bio-Rad 2-D standards were added for determination of approximate $M_r$ and p$I$. Aliquots containing 150 µg of protein for analytical purposes or 400 µg of protein for micropreparative separation, respectively, were precipitated overnight with 7.5 volumes of acetone containing 0.2 % (w/v) DTT at -20 °C. After washing the pellets again in the same solution, the samples were resolubilized in 350 µL of rehydration solution containing 7 M urea, 2 M thiourea, 1 % (w/v) C7BzO, 40 mM Tris-base, 70 mM DTT and 2 % (v/v) Pharmalyte 3/10 by incubating at 20 °C for 1 h. The samples were centrifuged again (16000 x g, 20 min, 15 °C) before loading on 18 cm nonlinear immobilized pH gradients (IPG) 3-10 (Bio-Rad, Hercules, CA) by in-gel rehydration.

Proteins were separated by isoelectric focusing using PROTEAN IEF Cell (Bio-Rad). The voltage was varied from 100 V (100 Vh, rapid), 500 V (500 Vh, linear), 1000 V (1000 Vh, linear) to 8000 V (95000 Vh, rapid), subsequently. The paper electrode wicks were changed 10 times during the first 10 kVh (the anodic wicks were soaked with water and the cathodic ones with 50 mM DTT). The IPGs were stored frozen at –80 °C.

The IPG strips containing total cell proteins were equilibrated for 12 min in a solution containing 6 M urea, 30 % (v/v) glycerol, 2 % (w/v) SDS, 50 mM Tris/HCl pH 8.8, trace of bromphenol blue and 1 % (w/v) DTT and then for a further 12 min in the same buffer except that DTT was replaced with 2.5 % (w/v) iodoacetamide. The IPGs were then embedded onto SDS-PAGE gels (20 cm x 20 cm in size, 1 mm thick) using 0.5 % (w/v) low-melting agarose in Laemmli electrode buffer. In the second dimension, homogenous (12 % T, 1.07 % C) SDS-PAGE gels, Laemmli buffer system (Laemmli 1970) and PROTEAN Plus Dodecacell were used. After an initial ramp up period of 2 h at 50 V, the gels were run at 100 V for about 20 h at 4 °C. The gel patterns were visualized by tetrathionate-silver nitrate staining (Rabilloud 1992) for analytical purposes or by SYPRO Ruby (Molecular Probes) in the case of micropreparative separations according to manufacturer's instructions. GS-800 and Pharos FX Pro instruments (Bio-Rad) were used for gel scanning. Spot detection, background subtraction, spot matching and data normalization using a local regression model method were performed using PDQUEST 8.0 software.

### 3.3 Statistical analysis of 2D-PAGE data

The normalized data exported from PDQUEST 8.0 were analyzed as follows: Values estimated by threshold level were excluded from the analysis. To reveal differences between groups, significance analysis of microarrays (SAM) (Tusher et al. 2001) was performed if there were at least 3 replicates in each of the compared groups. Proteins were considered as significantly differentially regulated if the false discovery rate (FDR) did not exceed 10 % and if the mean quantitative change was higher than 2 (up-regulation) or lower than 0.5 (down-regulation). In order to visualize the effect of selected proteins, hierarchical clustering based on Spearman correlation was performed. Data analysis was performed in a R-2.8.1 environment for statistical computing (R_Development_Core_Team 2008). For SAM the "samr" package was used, and clustering was performed using the package "cluster".

### 3.4 Mass spectrometry analyses

Sypro Ruby-stained protein spots selected for MS analysis were excised from 2D-PAGE gels. After destaining, the proteins in the gel pieces were incubated with trypsin (sequencing grade, Promega) at 37 °C for 2 h (Havlis et al. 2003). Peptide mass fingerprinting and tandem mass spectrometry (MS/MS) analyses were performed by matrix-assisted laser desorption-ionization mass spectrometry (MALDI-MS) with an Ultraflex III mass spectrometer (Bruker Daltonik, Bremen, Germany).

Sample preparation protocol for MALDI-MS employing α-cyano-4-hydroxycinnamic acid solution prepared according to Havlis (Havlis *et al.* 2003) used as the matrix in combination with AnchorChip target was used to enhance measurement sensitivity. The sample (1 µl) was mixed with matrix solution on the target in a 2:1 ratio. Peptide maps were acquired in reflectron positive mode (25 kV acceleration voltage) with 800 laser shots. Twelve dominant peaks within 700 – 3600 Da mass range and minimum S/N 10 were picked for MS/MS analysis employing laser induced dissociation – "LIFT" arrangement with 600 laser shots for each peptide. Known autoproteolytic products of trypsin were used for internal calibration of digested peptides. In absence of these products, an external calibration procedure was employed, using a mixture of seven peptide standards (Bruker Daltonik) covering the mass range of 1000 – 3100 Da. The Flex Analysis 3.0 and MS Biotools 3.1 (Bruker Daltonik) software were used for data processing.

In case of insignificant or negative results of the MS/MS ion search, tryptic digests were subjected to electrospray ionization liquid chromatography-tandem mass spectrometry (ESI-LC-MS/MS) analysis. LC-MS/MS experiments were accomplished on a high performance liquid chromatography system consisting of a gradient pump (Ultimate), autosampler (Famos) and column switching device (Switchos; LC Packings, Amsterdam, The Netherlands) on-line coupled with an HCTultra PTM Discovery System ion trap mass spectrometer (Bruker Daltonik). The column used for LC separation was filled according to a previously described procedure (Planeta *et al.* 2003). Prior to LC separation, tryptic digests were concentrated and desalted using PepMap C18 trapping column (300 µm x 5 mm, LC Packings). Sample volume was 15 µl. After washing with 0.1 % formic acid, the peptides were eluted from the trapping column using an acetonitrile/water gradient (4 µL/min) onto a fused-silica capillary column (320 µm x 180 mm), on which peptides were separated. This column was filled with 4-µm Jupiter Proteo sorbent (Phenomenex, Torrance, CA). The mobile phase A consisted of acetonitrile/0.1 % formic acid (5/95 v/v) mixture and the mobile phase B consisted of acetonitrile/0.1 % formic acid (80/20 v/v) mixture. The gradient elution started at 5 % of mobile phase B, and after 4 minutes, it was increased linearly from 5 % to 50 % during 55 minutes. The analytical column outlet was connected to the electrospray ion source via a 50-µm-inner diameter fused-silica capillary. Nitrogen was used as nebulizing as well as drying gas. The pressure of nebulizing gas was 15 psi. The temperature and flow rate of drying gas were set to 300 °C and 6 L/min, respectively, and the capillary voltage was 4.0 kV. The mass spectrometer was operated in the positive ion mode in an m/z range of 300 – 1500 for MS and 100 - 3000 for MS/MS scans. Extraction of the mass spectra from the chromatograms, mass annotation and deconvolution of the mass spectra were performed using DataAnalysis 4.0 software (Bruker Daltonik).

### 3.5 Mass spectrometry data processing

MASCOT 2.0 (MatrixScience, London, UK) search engine was used for processing the MS and MS/MS data. Database searches were done against the translated genome sequence data of *P. denitrificans* downloaded from http://genome.ornl.gov/microbial/pden/, the last sequence version released in 2006). A mass tolerance of up to 30 ppm was accepted during processing MALDI-MS data for PMF and 0.6 Da during processing laser-induced dissociation -"LIFT" data for MS/MS ion searches. For ESI-MS/MS data, mass tolerances of peptides and MS/MS fragments for MS/MS ion searches were 0.5 Da. Oxidation of methionine and carbamidomethylation of cysteine as optional and fixed modifications, respectively, and one enzyme miscleavage were set for all searches. Gene annotations are consistent with *P. denitrificans* genome database at http://genome.ornl.gov/microbial/pden/. Note: In the time of preparation of this chapter, the *Paracoccus denitrificans* genome database was just in the process of moving to a new address: http://genome.jgi-psf.org/parde/parde.download.ftp.html.

## 4. Web accessible 2D-PAGE dataset of *P. denitrificans* proteome

PDQUEST 8.0 software was used for image analysis of 2D-PAGE gels and handling and keeping all operational data in a local file. This software also served for gel calibration, spot numbering and quantitation. Local files in PDQUEST format contain 2-D maps of total cell lysates and membrane fractions annotated with spot numbers, experimental Mr/pI, gel ID numbers, identification status, MS identification mode, protein name and UniProt accession number.

These data have been submitted to the "Proteome Database System for Microbial Research at Max Planck Institute for Infection Biology" in Berlin, Germany where they are now stored in the "Proteome 2D-PAGE Database" subsection (http://www.mpiib-berlin.mpg.de/2D-PAGE/) (Mollenkopf *et al.* 1999). The "Proteome 2D-PAGE Database" currently contains 11146 protein identifications from 10975 spots and 3124 mass peaklists in 55 reference maps representing experiments from 26 different organisms and strains. The data were submitted by 104 submitters from 30 institutes from 13 nations. The aim of the PDBS is to share proteomics information in a readily manner with the scientific community as an invitation for data mining. Showing experimental data like MS peaklists and raw spectra leads to more transparency of the results. In addition, protein identification data are integrated with genomic, metabolic and other biological knowledge sources to increase the value of the primary data.

The frontend of the "Proteome 2D-PAGE Database", i.e. the website, is dynamically generated mainly by a combination of PERL and CGI, but also JAVA, PHP and R (http://www.r-project.org/) are used. The data in the backend are organized in a relational database under the control of MySQL (http://www.mysql.com) as database management system.

The user of the *P. denitrificans* 2D-PAGE dataset can find three 2-D maps of *P.denitrificans* proteome: (i) Coomassie-stained 2-D map of total cell lysate (Fig. 2, 26 proteins identified) , (ii) silver-stained map of membrane fraction (Fig. 3, 14 proteins identified) and (iii) Sypro Ruby-stained total cell lysate map (Fig. 4, 640 proteins identified). The third Sypro Ruby total cell lysate map was prepared with the latest technologies and protocols and contains the most comprehensive annotation, including the possibility of downloading the complete list of ORFs identified at proteome level.

The user of the database has several possibilities to highlight the MS analyzed spots according to MS analysis results: He/she can highlight (i) only identified spots (ii) only non-identified spots, (iii) all spots, and (iv) none spots. The spots with significant identification are marked with a red cross while the spots without identification are labeled with a blue cross. A zoom function is available; a detailed map view can be thus obtained. If the user moves the cursor over a cross, spot number and protein name appears. If he clicks on the

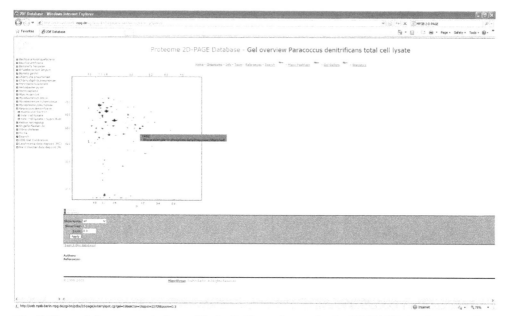

Fig. 5. The web browser window of *P. denitrificans* dataset in the 2D-PAGE database: total cell lysate 2D-PAGE map. The cursor-selected spot is annotated with its protein name.

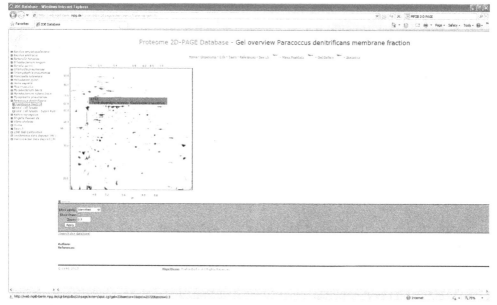

Fig. 6. The window of *P. denitrificans* dataset in the 2D-PAGE database: 2D-PAGE map of the membrane fraction.

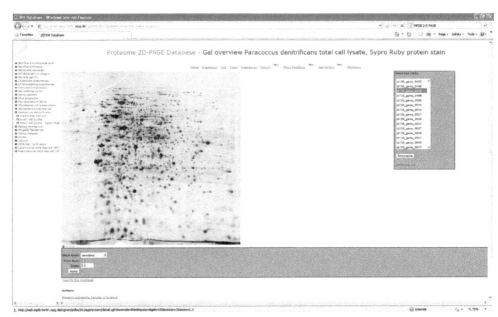

Fig. 7. Sypro Ruby-stained total cell lysate map gel containing 640 identified proteins. The list of corresponding open reading frames identified by MS can be downloaded directly from this web page.

protein spot, the page with more detailed protein information is opened, including spot molecular weight, spot pI, UniProt accession number, gene locus, protein name, identification method, identification status and sequence coverage. If the visitor is interested in protein amount changes among different growth conditions, he can get detailed information in publicly accessible supplementary tables related to each published project.

Originally, only 8 proteins were identified by peptide mass fingerprinting from 49 analyzed spots within the first Coomassie-stained total cell lysate map and in membrane fraction map due to unknown sequence of *P. denitrificans* genome (Bouchal *et al.* 2004). Subsequent genome sequencing at Joint Genome Institute (http://www.jgi.doe.gov) resulted in public release of a *P. denitrificans* complete genome sequence in 2006. Thereafter the translated sequence data were transformed in the file applicable for MASCOT searches. Remaining 41 proteins were successfully identified using available MS spectra and the database gels were updated by new identifications. This substantial update underlines the significance of genome information on the effectivity of protein identification in proteomics.

### 5. *P. denitrificans* -omics projects facilitated *via* web accessible 2D-PAGE database

Because of the current progress in –omics methods and applications, it is expected that other laboratories interested in *P. denitrificans* biology will implement proteomic approach into their method toolboxes. The *P. denitrificans* web accessible dataset can help them with formulating their hypotheses, planning their experiments, orientation in their own proteome maps as well as facilitating communication among laboratories.

The perspective results coming from future studies on *P. denitrificans* are likely to be important from the bioenergetic point of view, providing a basis for understanding the well-known nutritional versatility of this bacterium. Since the rate of metabolic processes, in many cases depending on the substrate used, is often related to levels of involved proteins, identification of pivotal metabolic enzymes is one of the first tasks of proteomic research. From this point of view, the first important finding among the published results is the identification of proteins involved in denitrification. Nitrate reductase β-subunit, nitrite reductase and nitrous oxide reductase are the key enzymes of denitrification pathway and all of them were detected using proteomic approach in Sypro Ruby total cell lysate proteome maps as spots 5710 (nitrate reductase β-subunit), 1701 (nitrite reductase) and 2701 (nitrous oxide reductase).

Since the synthesis of denitrification enzymes is regulated by FNR-type transcription regulators (FnrP, NNR, NarR), looking for spot position of these regulatory proteins in 2-D maps is a next logical step of the expert user. Although their level is probably too low for detection on 2-D PAGE gels, an *UspA* gene product (gene 1849) as a direct neighbour of *fnrP* in the genome was detected in spot 3211 (Sypro Ruby 2-D map). Its expression profile was similar to *fnrP* gene product, suggesting their co-expression in single operone (Bouchal *et al.* 2010). Terminal oxidases (*aa₃*, *cbb₃* and *ba₃* types in *P. denitrificans*, Fig. 4) are very hydrophobic proteins as most of their subunits contain more than one transmembrane domain (information obtained using PSORT algorithm, http://psort.nibb.ac.jp) and their identification on 2-D PAGE gels cannot be expected - for review, see (Santoni *et al.* 2000). On the other hand, we identified α, β and ε-subunits of $F_0F_1$-ATPase (spot 2103 in Coomassie-stained total cell lysate map and spots 0011, 0120, 0503, 1204, 1319, 5529 in a map of membrane fraction) being in many cases downregulated under anaerobic growth conditions as a probable result of general slowing down of the energetic metabolism. Among the proteins induced with azide (Bouchal *et al.* 2004; Bouchal *et al.* 2011) Fe/Mn superoxide dismutase (spot 6106 in Coomassie-stained total cell lysate map and spot 5001 in a map of membrane fraction) was identified, indicating generation of an increased amount of reactive oxygen species, possibly as a result of the increased degree of reduction of respiratory components. This was independently confirmed at transcript and enzyme activity level (Bouchal *et al.* 2011). Furthermore, synthesis of Fe/Mn superoxide dismutase is independent of FNR-type regulators (Bouchal *et al.* 2010). The only protein in membrane fraction induced synergically by nitrate and azide (spot 1702 in membrane fraction map) was TonB dependent receptor, a protein involved in iron transport. His ORF is located very close to preudoazurin gene, an alternative electron-transporter in a denitrification pathway. We also found glyceraldehyde-3-phosphate dehydrogenase (spot 7402 in Coommassie-stained total cell lysate map) non-affected with different growth conditions, so it could serve as an internal standard for gene expression comparison.

Our subsequent comprehensive study focused on FNR-type transcription regulators (Bouchal *et al.* 2010) revealed four significant protein clusters according to correlation of their levels under aerobic, semiaerobic and semiaerobic with nitrate growth conditions (see Fig. 8, spot numbers correspond to Sypro Ruby-stained proteome map): (i) The first cluster contains proteins involved in the FnrP regulon. It involves nitrous oxide reductase (spot 2701), UspA protein (spot 3211), and two OmpW proteins (spots 4107 and 5105) as well as two spots 501 and 8701 identified as unknown proteins. The direct regulation of nitrous oxide reductase, UspA and OmpW proteins by FnrP is a new finding from the mentioned study. (ii) Second cluster involves proteins regulated *via* additional regulators, including

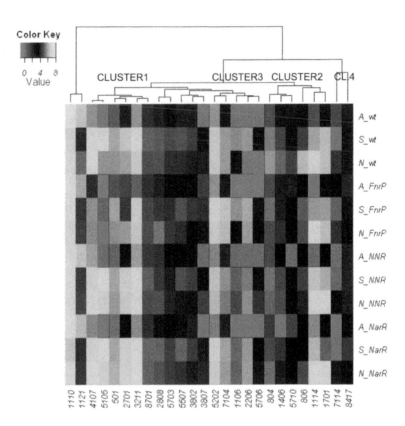

Fig. 8. The Spearman correlated clustering of the regulation profiles of proteins differentially regulated in response to mutation in FnrP, NNR and NarR transcription regulators and/or to oxygen and nitrate. See (Bouchal *et al.* 2010) for details.

proteins involved in NNR and NarR regulons. This cluster contains two TonB dependent receptors (spots 804, 806 and 1814), nitrate reductase β-subunit (spot 5710), a TenA-type transcription regulator (spot 1114), nitrite reductase (spot 1701) and an unknown protein with an alpha/beta hydrolase fold (spot 1406). The clustering of the TenA transcription regulator with nitrite reductase might well be indicative for the involvement of such an additional regulator. This clustering also indicates that ranking of the above mentioned proteins specifically under the NNR or the NarR regulon is less straightforward, probably since both regulators are activated by the reduction products of a common substrate, nitrate. (iii) The third cluster involves proteins whose amount is affected by the growth condition rather than by mutations in the FNR-type proteins. As such, these proteins may be part of a more global regulatory switch. This cluster contains SSU ribosomal protein S305 / σ54 modulation protein (spot 5202) and two SDR proteins (spots 2206 and 7104). (iv) The fourth cluster contains only the proteins specifically upregulated in cells grown semiaerobically in the presence of nitrate: one uncharacterized protein (spot 7114) and an ABC-type transporter of unknown function (spot 8417) (Bouchal *et al.* 2010).

## 6. Perspectives

In the upcoming time, development of the *P. denitrificans* 2-D PAGE dataset can continue in the following ways. (1) The identification of gel proteins spots not yet analyzed *via* mass spectrometry will continue, followed by immediate updates of the dataset. Resolution at the protein species level as obtained by 2D-PAGE-MS methods has in contrast to the bottom-up LC-MS approach the advantage to consider protein speciation (Jungblut *et al.* 2008) and will allow the analysis of protein species-specific regulation as already described within the phosphoproteome of *Helicobacter pylori* infected human stomach adenocarcinoma AGS cell line (Holland *et al.* 2011). (2) Implementation of new perspective analytical methods is in progress. 2DLC-MS/MS-based approaches with stable isotope labelling (iTRAQ, SILAC), or with label-free quantification can serve as a method complementary to 2D-PAGE-MS. These approaches are also helpful when identification of integral membrane proteins with non-membrane domains is required (Wu *et al.* 2003; Bouchal *et al.* 2009). (3) In the case of the most hydrophobic or low abundant proteins, non-proteomic approaches like qRT-PCR and cDNA chips are available for the study of gene expression. (4) Further progress of *P. denitrificans* genome information and annotation is expected. Direct accession from the *P. denitrificans* dataset in the 2D-PAGE database into *P. denitrificans* genome database would underline the integrity of genomic and proteomic data and facilitate finding the relations between protein level, gene location and gene function.

The data collection in the similar format raises the possibility of data comparison between different proteomes. During reading the communications about various microbial proteomic studies, one can be surprised how many identical (or very relative) proteins have been identified in different organisms, while other physiologically important proteins may be underrepresented in 2-D gels. It is more interesting if the number of theoretical ORFs in genomes is taken into account. It is obvious that protein hydrophobicity and solubility together with copy number plays an important role in these observations (Wilkins *et al.* 1998; Santoni *et al.* 2000; Jungblut *et al.* 2010). Using an integrated "Proteome 2D-PAGE Database" (http://www.mpiib-berlin.mpg.de/2D-PAGE/) covering a number of bacterial proteomes, it is easier to predict whether the protein of one's interest will, or will not be identified using proteomics approach and can be quantified this way. Such a tool is very useful for people making a choice of the best method for a screening of the gene(s) expression, or the protein(s) synthesis in *Paracoccus denitrificans* focused projects.

## 7. Conclusions

We feel that -omics approach is a powerful tool for a study of thousands of cellular genes and proteins in *P. denitrificans* and their variability between different growth conditions. Keeping in mind the principles of proteomics, it can be viewed as a screening tool able to reveal the specific changes, the detection of which would require significantly larger amount of work using classical approaches. With regard to nature of various environmental effects, „omics" approaches itself cannot provide final evidences for their mechanisms. However, the hypotheses obtained using this toolbox can provide a firm bases for targeted functional studies using integrated modern biochemical, bioinformatic and molecular-biological methodologies. We hope that our dataset within the 2D-PAGE database will be useful in designing such integrated *P. denitrificans* projects and in facilitating the international interlaboratory cooperation.

## 8. Acknowledgements

The authors would like to thank Dr. Rob van Spanning, Vrije Universiteit Amsterdam, the Netherlands, for providing the bacterial strains and Dr. Eva Budinská, Masaryk University Brno, Czech Republic & Swiss Institute for Bioinformatics, Lausanne, Switzerland, for statistical analyses. The experiments and chapter publishing costs were covered by Czech Ministry of Education (project No. MSM0021622413). P.B. was partly supported by European Regional Development Fund and the State Budget of the Czech Republic (RECAMO; CZ 1.05/2.1.00/03.0101) and by Czech Ministry of Health (MZ0MOU2005), Z.Z. was supported by the project "CEITEC - Central European Institute of Technology" (CZ.1.05/1.1.00/02.0068) from European Regional Development Fund.

## 9. References

Bouchal, P. and I. Kucera (2002) Two-dimensional electrophoresis in proteomics: Principles and applications. *Chemicke Listy*. 97, 1, 29-36, ISSN 0009-2770

Bouchal, P. and I. Kucera (2004) Examination of membrane protein expression in Paracoccus denitrificans by two-dimensional gel electrophoresis. *Journal of Basic Microbiology*. 44, 1, 17-22, ISSN 0233-111X

Bouchal, P., P. Precechtelova, Z. Zdrahal and I. Kucera (2004) Protein composition of *Paracoccus denitrificans* cells grown on various electron acceptors and in the presence of azide. *Proteomics*. 4, 9, 2662-2671, ISSN 1615-9853

Bouchal, P., T. Roumeliotis, R. Hrstka, R. Nenutil, B. Vojtesek and S. D. Garbis (2009) Biomarker discovery in low-grade breast cancer using isobaric stable isotope tags and two-dimensional liquid chromatography-tandem mass spectrometry (iTRAQ-2DLC-MS/MS) based quantitative proteomic analysis. *Journal of Proteome Research*. 8, 1, 362-373, ISSN 1535-3893 (Print)

Bouchal, P., I. Struharova, E. Budinska, O. Sedo, T. Vyhlidalova, Z. Zdrahal, R. van Spanning and I. Kucera (2010) Unraveling an FNR based regulatory circuit in *Paracoccus denitrificans* using a proteomics-based approach. *Biochimica Et Biophysica Acta-Proteins and Proteomics*. 1804, 6, 1350-1358, ISSN 1570-9639

Bouchal, P., T. Vyhlidalova, I. Struharova, Z. Zdrahal and I. Kucera (2011) Fe/Mn superoxide dismutase-encoding gene in *Paracoccus denitrificans* is induced by azide and expressed independently of the FNR-type regulators. *Folia Microbiologica*. 56, 1, 13-17, ISSN 1874-9356 (Electronic) 0015-5632 (Linking)

Burnell, J. N., P. John and F. R. Whatley (1975) The reversibility of active sulphate transport in membrane vesicles of Paracoccus denitrificans. *Biochemical Journal*. 150, 3, 527-536, ISSN 0264-6021 (Print) 0264-6021 (Linking)

Havlis, J., H. Thomas, M. Sebela and A. Shevchenko (2003) Fast-response proteomics by accelerated in-gel digestion of proteins. *Analytical Chemistry*. 75, 6, 1300-1316, ISSN 0003-2700 (Print)

Holland, C., M. Schmid, U. Zimny-Arndt, J. Rohloff, R. Stein, P. R. Jungblut and T. F. Meyer (2011). Quantitative phosphoproteomics reveals link between *Helicobacter pylori* infection and RNA splicing modulation in host cells. *Proteomics*. 11, 14, 2798-2811, ISSN 1615-9861 (Electronic) 1615-9853 (Linking)

Jungblut, P. R., H. G. Holzhutter, R. Apweiler and H. Schluter (2008) The speciation of the proteome. *Chemistry Central Journal.* 2, 16, ISSN 1752-153X (Electronic) 1752-153X (Linking)

Jungblut, P. R., F. Schiele, U. Zimny-Arndt, R. Ackermann, M. Schmid, S. Lange, R. Stein and K. P. Pleissner (2010) Helicobacter pylori proteomics by 2-DE/MS, 1-DE-LC/MS and functional data mining. *Proteomics.* 10, 2, 182-193, ISSN 1615-9861 (Electronic) 1615-9853 (Linking)

Koutny, M. and I. Kucera (1999) Kinetic analysis of substrate inhibition in nitric oxide reductase of Paracoccus denitrificans. *Biochemical and Biophysical Research Communications.* 262, 2, 562-564, ISSN 0006-291X (Print)

Laemmli, U. K. (1970) Cleavage of structural proteins during the assembly of the head of bacteriophage T4. *Nature.* 227, 5259, 680-685, ISSN 0028-0836 (Print) 0028-0836 (Linking)

Mollenkopf, H. J., P. R. Jungblut, B. Raupach, J. Mattow, S. Lamer, U. Zimny-Arndt, U. E. Schaible and S. H. Kaufmann (1999) A dynamic two-dimensional polyacrylamide gel electrophoresis database: the mycobacterial proteome via Internet. *Electrophoresis.* 20, 11, 2172-2180, ISSN 0173-0835 (Print) 0173-0835 (Linking)

Planeta, J., P. Karasek and J. Vejrosta (2003) Development of packed capillary columns using carbon dioxide slurries. *Journal of Separation Science.* 26, 6-7, 525-530, ISSN 1615-9306 (Print), 1615-9314 (Online)

R_Development_Core_Team (2008) A language and environment for statistical computing, R Foundation for Statistical Computing Vienna, Austria

Rabilloud, T. (1992) A comparison between low background silver diammine and silver nitrate protein stains. *Electrophoresis.* 13, 7, 429-439, ISSN 0173-0835 (print), 1522-2683 (online)

Santoni, V., M. Molloy and T. Rabilloud (2000) Membrane proteins and proteomics: un amour impossible? *Electrophoresis.* 21, 6, 1054-1070, ISSN 0173-0835 (Print) 0173-0835 (Linking)

Tusher, V. G., R. Tibshirani and G. Chu (2001) Significance analysis of microarrays applied to the ionizing radiation response. *Proceedings of National Academy of Sciences U S A.* 98, 9, 5116-5121, ISSN 0027-8424 (Print)

Van Spanning, R. J., A. P. De Boer, W. N. Reijnders, S. Spiro, H. V. Westerhoff, A. H. Stouthamer and J. Van der Oost (1995) Nitrite and nitric oxide reduction in *Paracoccus denitrificans* is under the control of NNR, a regulatory protein that belongs to the FNR family of transcriptional activators. *FEBS Letters.* 360, 2, 151-154, ISSN 0014-5793 (Print)

Van Spanning, R. J. M., A. P. N. De Boer, W. N. M. Reijnders, H. V. Westerhoof, A. H. Stouthamer and J. van der Oost (1997) FnrP and NNR of *Paracoccus denitrificans* are both members of the FNR family of transcriptional activators but have distinct roles in respiratory adaptation in response to oxygen limitation. *Molecular Microbiology.* 12, 5, 893-907, ISSN 0950-382X (Print), 1365-2958 (Online)

Veldman, R., W. N. M. Reijnders and R. J. M. van Spanning (2006) Specificity of FNR-type regulators in *Paracoccus denitrificans*. *Biochemical Society Transactions.* 34, 1, 94-96, ISSN 0300-5127 (Print), 1470-8752 (Electronic)

Wilkins, M. R., E. Gasteiger, L. Tonella, K. Ou, M. Tyler, J. C. Sanchez, A. A. Gooley, B. J. Walsh, A. Bairoch, R. D. Appel, K. L. Williams and D. F. Hochstrasser (1998)

Protein identification with N and C-terminal sequence tags in proteome projects. *Journal of Molecular Biology.* 278, 3, 599-608, ISSN 0022-2836 (Print) 0022-2836 (Linking)

Wood, N. J., T. Alizadeh, S. Bennett, J. Pearce, S. J. Ferguson, D. J. Richardson and J. W. Moir (2001) Maximal expression of membrane-bound nitrate reductase in *Paracoccus* is induced by nitrate via a third FNR-like regulator named NarR. *Journal of Bacteriology.* 183, 12, 3606-3613, ISSN 0021-9093 (Print), 1098-5530 (Online)

Wu, C. C., M. J. MacCoss, K. E. Howell and J. R. Yates, 3rd (2003) A method for the comprehensive proteomic analysis of membrane proteins. *Nature Biotechnology.* 21, 5, 532-5l8, ISSN 1087-0156 (Print) 1087-0156 (Linking)

Zumft, W. G. (1997) Cell biology and molecular basis of denitrification. *Microbiology and Molecular Biology Reviews.* 61, 533-616, ISSN 1092-2172 (Print), 1098-5557 (Online)

# 8

# Preparation of Protein Samples for 2-DE from Different Cotton Tissues

Chengjian Xie[1], Xiaowen Wang[2], Anping Sui[2*] and Xingyong Yang[1,2*]
*[1]Chongqing Normal University, Chongqing*
*[2]Southwest University, Chongqing*
*China*

## 1. Introduction

Cotton today is one of the most important economic crops. Cotton fiber is the most used material for the textile industry, and takes an important strategic status in the world economy. Proteomics is one of the most important techniques in the post-genome era, and two-dimensional electrophoresis (2-DE) is one key technology for proteomics. Protein extraction and sample preparation are of prime importance for optimal 2-DE results (Isaacson et al., 2006). The cotton is a highly recalcitrant plant material and rich in compounds such as polysaccharides, polyphenols, nucleic acids, cellulose, and other secondary metabolites, which interfere with protein extraction, produce highly diluted protein extracts, and affect protein migration in 2-DE (Görg et al., 2000). Moreover, modified or different protein extraction method should be used for different cotton tissues (e.g., leaves, roots, seeds, and stems), which contain different secondary metabolites.

Here, several protocols to extract total proteins for different cotton tissues are described based on methods routinely used in our laboratory.

## 2. Materials

Ultrapure water (doubly distilled, deionized, > 18 MΩ) is used for all reagent preparation. Reagent grades should be of the highest quality.

1.  1 M Tris-saturated phenol (pH 8.0)
2.  Extraction buffer: 0.1 M Tris-HCl, pH 8.0, containing 30% w/v sucrose, 2% w/v SDS, 1 mM phenylmethanesulfonyl fluoride (PMSF), 2% v/v thioglycol 2-mercapitoethanol (2-ME).
3.  Lysis buffer: 7 M urea, 2 M thiourea, 4% w/v CHAPS, 65 mM DTT, and 0.5% v/v carrier ampholytes.
4.  Equilibration buffer: 6 M urea, 20% w/v glycerol, 2% w/v SDS, and 50 mM Tris-HCl, pH 8.8.
5.  Staining solution: 0.12% w/v Coomassie brilliant blue G-250, 10% w/v ammonium sulfate, 10% w/v phosphoric acid, 20% v/v methanol.

---

*Corresponding Authors

## 3. Methods

All plant tissue samples should be ground to fine powder with a pre-chilled mortar and pestle in liquid nitrogen. Before grinding, silicon dioxide (SiO$_2$) and polyvinylpolypyrrolidone (PVPP, 10% w/w of sample weight) were added into mortar. The finely ground powder (ca. 0.5 g per tube) was immediately transferred into a 10-mL centrifuge tube (see Note 1) precooled in liquid nitrogen. The powder sample can be immediately used or stored in a -80 °C freezer until protein extraction.

### 3.1 Extraction from leaves
- The powder sample was resuspended in 4 mL 10% v/v trichloroacetic acid (TCA) in acetone (see Note 2) and extensively homogenized.
- Centrifuge (12, 000$g$ at 4 °C) for 5 min, the pellet of proteins was washed once with 5 mL 0.1 M ammonium acetate in 80% v/v methanol and once with cold 80% v/v acetone.
- The pellet was dried in vacuum (see Note 3).
- It was resuspended in 3 mL of extraction buffer.
- An equal volume of 1 M Tris-saturated phenol (pH 8.0) was added, and then the mixture was homogenized on ice for 5 min. The upper phenol phase was collected after centrifuge (12, 000$g$ at 4 °C) for 5 min, and this extraction step was repeated once (see Note 4).
- The total phenol phase was transferred into a new tube, and an equal volume of extraction buffer was added into the total phenol phase, and then the mixture was homogenized on ice for 5 min. The upper phenol phase was collected after centrifuge (12, 000$g$ at 4 °C) for 5 min (see Note 4).
- The proteins were precipitated with five volumes of 0.1 M ammonium acetate in methanol overnight at -20 °C.
- Centrifuge (12, 000$g$ at 4 °C) for 10 min. The collected protein pellets were washed once with 3 mL methanol, and then washed once with 3 mL cold 80% v/v acetone in water. The pellets were dried in a freeze vacuum dryer for 10 min and stored at -80 °C.

### 3.2 Extraction from roots (see Note 5)
- The powder sample was resuspended in 3 mL of extraction buffer.
- An equal volume of 1 M Tris-saturated phenol (pH 8.0) was added, and then the mixture was homogenized on ice for 5 min. The upper phenol phase was collected after centrifuge (12, 000$g$ at 4°C) for 5 min, and this extraction step was repeated once (see Note 4).
- The total phenol phase was transferred into a new tube, and an equal volume of extraction buffer was added into it. The mixture was homogenized on ice for 5 min, and the upper phenol phase was collected after centrifuge (12, 000$g$ at 4°C) for 5 min (see Note 4).
- Precipitation the phenol phase with 5 volumes of 0.1 M ammonium acetate in methanol overnight at -20°C.
- Centrifuge (12, 000 g at 4°C) for 10 min. The protein pellets were washed once with 3 mL methanol, and then washed once with 3 mL 80% v/v acetone in water. The pellets were dried in a freeze vacuum dryer for 10 min and stored at -80 °C.

### 3.3 Extraction from fibers (see Note 6)

- The ground powder was resuspended in 4 mL acetone and extensively homogenized. The sample was kept at -20 °C overnight.
- Centrifuge (8, 000g at 4 °C) for 5 min (see Note 6). Discard the supernatant and the pellets were washed once with 4 mL acetone containing 2% v/v 2-ME.
- Centrifuge (10, 000 g at 4°C) for 5 min, and the wash was repeated once.
- Discard the supernatant and the pellet was dried in vacuum.
- Resuspend the pellet with 4 mL of extraction buffer.
- An equal volume of 1 M Tris-saturated phenol (pH 8.0) was added and homogenized on ice for 5 min. Centrifuge (12, 000g at 4 °C) for 5 min.
- The upper phenol phase was collected. The phenol extraction procedure was repeated once.
- The collected phenol phase was precipitated with 5 volumes of 0.1 M ammonium acetate in methanol overnight at -20 °C.
- Centrifuge (10, 000g at 4 °C) for 10 min. Discard the supernatant and the pellet was washed twice with cold 0.1 M ammonium acetate in methanol. Wash twice with cold 80% acetone in water. The pellet was dried in a freeze vacuum dryer and stored at -80 °C.

### 3.4 Proteins pellet resuspension

Proteins pellet was resuspended in lysis buffer and shaked for 1 h (IKA Vortex Genius 3, Staufen, Germany). After centrifugation at 15, 000g for 20 min to remove debris, the supernatant could be used immediately for first-dimensional IEF gels. Protein concentration was determined using the Bradford method (Bradford, 1976) with bovine serum albumin as a standard.

### 3.5 Two-dimensional gel electrophoresis

The 2-D gel electrophoresis (2-DE) protocol was adapted by O'Farrell (1975). The first electrophoresis was performed using immobilized pH gradient (IPG) strips on an IPGphor isoelectric focusing (IEF) system (Amersham Pharmacia, San Francisco, CA). For example in our experiment, the IPG strips (13 cm, 3–10 nonlinear pH gradient; GE Healthcare, Piscataway, NJ) were rehydrated with 250 µl of rehydration buffer (containing 370 µg proteins). Focusing was then performed at 20 °C as follows: active rehydration at 30 V for 12 h, 200 V for 2 h, 500 V for 3 h, 1,000 V for 4 h, 8,000 V for 5 h, with a gradient increase in voltage between 8,000 V and 40,000 V. After IEF, the proteins in the strips were reduced with 1% w/v DTT in 10 ml of equilibration buffer for 15 min and alkylated with 2.5% w/v iodoacetamide in 10 ml of equilibration buffer for 15 min. The strips were transferred onto vertical 10.5% w/v SDS-PAGE selfcast gels. The second electrophoresis (SDS-PAGE) was performed on an Amersham Hoefer SE 600 system (Amersham Pharmacia) at 10 mA for 1 h and 20 mA for 6 h at 15 °C.

### 3.6 Protein visualization

The 2-DE gel was stained with blue silver (Candiano et al., 1975). The gel was fixed in a solution of 40% v/v methanol and 10% v/v acetic acid for 30 min, washed in distilled water 4 times for 15 min, and finally incubated in a staining solution (see Note 7) overnight with gentle shaking. The gel was decolorized in distilled water.

As shown in Figure 1, a great many of protein spots were detected on the 2-DE image. The crude protein yield and the number of protein on 2-DE gels are also summarized in Table 1.

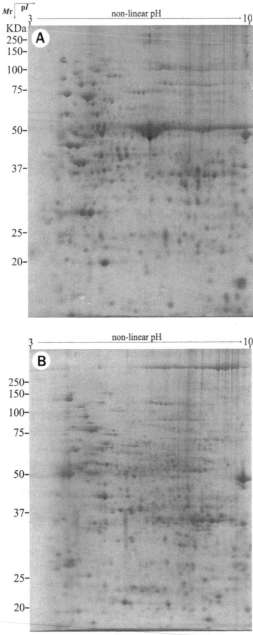

Fig. 1. Two-dimensional gel electrophoresis gel of proteins extracted from cotton leaves (A) and roots (B). Proteins (370 μg) were separated on a 13-cm pH 3–10 nonlinear gradient immobilized pH gradient strip and on 10.5% SDS-PAGE gel. The gels were stained using the blue-silver method. Mr, molecular mass; pI, isoelectric point.

| Tissues | Protein yield (mg/g; crude protein/powdered tissue) | The number of protein on 2-DE gels |
|---------|------------------------------------------------------|-------------------------------------|
| Leaves | 8.7 | About 900 |
| Roots | 5.1 | About 830 |
| Fibers | 4.6 | About 850 |

Table 1. The crude protein yield and the number of protein on 2-DE gels for different extraction methods. Proteins (370 µg) were separation on 13cm pH 3-10 non-linear gradient IPG strip and 10% SDS-PAGE gel. The gel was stained using Blue silver [4]. The above results are from different experiments and represent initial results. It can provide preliminary reference for selecting extraction method.

These results demonstrate that above protein extraction methods could be compatible with cotton different tissues.

## 4. Notes

1. Small plant samples can yield a sufficient amount of protein for 2-DE. Furthermore, using small plant samples easily extract the high purity of protein.
2. These mentioned solutions except for lysis buffer were stored at 4 °C and cooled on ice in advance for use.
3. A good principle for drying sample is that the edge of protein pellets turned white, and too dry samples is bad for resuspension.
4. To reduce loss of protein, this phenol extraction step was repeated once. However, some water soluble organic (such as polysaccharides and nucleic acids) was involved in collected total phenol phase and resulted in horizontal streaking visible on two dimensional gels. Once more wash step with an equal volume of extraction buffer was performed to remove water soluble organic in phenol phase.
5. TCA-acetone precipitation can effectively remove pigments (Xie et al., 1975). There are not too many pigments such as chlorophyll in cotton roots, so we could get the high purity protein samples if step 1-3 of "**Extraction from leaves**" was removed.
6. The method described here was adapted from published paper with minor modifications. Centrifugation at lower speeds (step 2) is beneficial to resuspend protein pellets. It, sometimes, is the necessary to use auxiliary tool for scraping pellets. Using TCA-acetone instead of acetone (step 1) is helpful for pellets' resuspension, although TCA-acetone precipitation could increase the loss of protein (Görg et al., 2004).
7. To prepare blue silver staining solution, phosphoric acid and distilled water (10% of the final volume) firstly were mixed, and then ammonium sulfate fine powder was added and completely dissolved on magnetic stirrer. Next, 0.12% (w/v) coomassie brilliant blue G-250 was added and stirred at least 2 hour. Fill to 80% final volume with distilled water, sequentially add 20% volume methanol and mix thoughly. These undissolved particles in staining solution were not needed to filter out and they can increase sensitivity of staining solution.

## 5. Acknowledgments

"Extraction from leaves" was adapted from Wang et al. method with modifications (Wang et al., 2003, 2006) and "Extraction from fibers" was based on the method previously described by Yao et al. (2006), with slight modifications. Furthermore, this work was supported by the National Natural Science Foundation of China (Grant no. 30771388), the Natural Science Foundation Project of CQ CSTC (Grant no. 2009BB1123) and the Fundamental Research Funds for the Central Universities (Grant no. XDJK2011C016).

## 6. References

Bradford, M.M. (1976) A rapid and sensitive method for the quantitation of microgram quantities of protein utilizing the principle of protein-dye binding. *Analytical Biochemistry*, Vol. 72, No. 1-2, (May 7), pp. 248-254, ISSN 0003-2697.

Candiano, G.; Bruschi, M.; Musante, L.; Santucci, L.; Ghiggeri, G.M.; Carnemolla, B.; Orecchia, P.; Zardi, L.; Righetti, P.G. (2004) Blue silver: A very sensitive colloidal Coomassie G-250 staining for proteome analysis. *Electrophoresis*, Vol.25, No.9, (May 15), pp.1327-1333, ISSN 0173-0835.

Görg, A,; Obermaier, C.; Boguth, G.; Harder, A.; Scheibe, B.; Wildgruber, R.; Weiss, W. (2000). The current state of two-dimensional electrophoresis with immobilized pH gradients. *Electrophoresis*, Vol. 21, No.6, (April 1), pp. 1037-1053, ISSN 0173-0835.

Görg, A.; Weiss, W.; Dunn, MJ. (2004) Current two-dimensional electrophoresis technology for proteomics. *Proteomics*, Vol. 4, No. 12, (December 20), pp. 3665-3685, ISSN 1615-9853.

Isaacson, T.; Damasceno, C.M.B.; Saravanan, R.S.; He, Y.; Catala, C.; Saladie, M.; Rose, J.K.C. (2006). Sample extraction techniques for enhanced proteomic analysis of plant tissues. *Nature Protocols*, Vol.1, No.2, (July 13), pp. 769-774, ISSN 1754-2189

O'Farrell, P.H. (1975) High resolution two-dimensional electrophoresis of proteins. *Journal of Biology Chemistry*, Vol.250, No.10, (May 25), pp. 4007-4021, ISSN 0021-9258

Wang, W.; Scali, M.; Vignani, R.; Spadafora, A.; Sensi, E.; Mazzuca, S.; Cresti, M. (2003) Protein extraction for two-dimensional electrophoresis from olive leaf, a plant tissue containing high levels of interfering compounds. *Electrophoresis*, Vol. 24, No.14, (July 15), pp. 2369-2375, ISSN 0173-0835.

Wang, W.; Vignani, R.; Scali, M.; Cresti, M. (2006) A universal and rapid protocol for protein extraction from recalcitrant plant tissues for proteomic analysis. *Electrophoresis*, Vol. 27, No.13, (July 1), pp. 2782-2786, ISSN 0173-0835.

Xie, C.; Wang, D.; Yang, X. (2009) Protein Extraction Methods Compatible with Proteomic Analysis for the Cotton Seedling. *Crop Science*, Vol. 49, No.2, (March 15), pp. 395-402, ISSN 1435-0653.

Yao, Y.; Yang, Y.; Liu, J. (2006) An efficient protein preparation for proteomic analysis of developing cotton fibers by 2-DE. *Electrophoresis*, Vol. 27, No. 22, (November 15), pp. 4559-4569, ISSN 0173-0835.

# Part 4

# Structural Proteomics

# 9

# The Utility of Mass Spectrometry Based Structural Proteomics in Biopharmaceutical Biologics Development

Parminder Kaur and Mark R. Chance
*Center for Proteomics and Bioinformatics,*
*Case Western Reserve University, Cleveland, OH*
*NeoProteomics, Inc., Cleveland, OH*
*USA*

## 1. Introduction

Proteins are essential components of living organisms and participate in nearly every biochemical process within cells. Examples of the processes include enzyme catalysis, cell signalling, host defense, metabolism, etc. These large, complex bio-molecules are connected by long chains of amino acids, that fold in very intricate patterns giving rise to a unique three-dimensional conformation. The biological function and physicochemical properties of a protein are determined by this higher order structure Ecroyd & Carver (2008); Hegyi & Gerstein (1999); Sadowski & Jones (2009). Most proteins tend to achieve the lowest possible free energy of the polypeptide chain and the surrounding solvent forming a native structure under physiological conditions Anfinsen (1973). This tightly folded conformation typically represents the biologically active state necessary for performing the required biochemical task. However, these macromolecules also have a temporal behavior leading to significant flexibility and dynamic motion because of the fluctuations in the surrounding electrostatic forces and hydrogen bonds that are important for maintaining conformations Henzler-Wildman & Kern (2007); Teilum et al. (2009). These temporal variations are important for certain functions such as protein-protein interactions and protein stability Kamerzell & Middaugh (2008); Travaglini-Allocatelli et al. (2009); van der Kamp et al. (2010). Thus, it is the presence of both spatial and temporal characteristics that allows for an ensemble of various molecular conformations to exist in solution. Change in environment of proteins, such as solvent acidity, urea concentration, temperature fluctuations, can change the folding pattern of the protein. Studying these partially or fully denatured states provides insights for understanding a variety of *in vivo* processes such as structural changes associated with aggregation, signal transduction, and transportation across membranes. Certain biological conditions can cause misfolding and aggregation of proteins, often causing severe disorders such as Alzheimer's disease, spongiform encephalopathies, and certain forms of diabetes Dobson (2003). Many genetic diseases are caused by protein-folding disorders, because an altered gene results in a modified protein sequence which is not able to undergo native folding and results in the disease phenotype Dobson (2001). Proteins have the ability to interact with one another, and can also bind to smaller ligands, which forms the basis of signaling and regulatory processes, playing a critical role in the mechanisms of drug activity. Owing to the critical importance of structure-function paradigm, the pioneering scientists, Anfinsen and Stein and Moore, who established the relationship between protein structure and function, were awarded the Nobel Prize in Chemistry in 1972 Anfinsen (1973); Moore & Stein (1973). The static protein structures commonly seen in X-ray images from the literature depict only one of the many possible conformations that the protein assumes at a particular instant in time. In fact, conformational dynamics are essential for mediating multifaceted functional roles performed by many proteins Fenimore et al. (2002); Frauenfelder et al. (1991); Huang & Montelione (2005). One particular example is the case of enzymes that require induced-fit binding behavior for proper operation Falke (2002); Schulz (1992). This suggests that the structural rearrangements of of such proteins represents a well balanced compromise between a highly ordered core conformation to ensure specificity, and a relatively flexible and dynamic state that maintains diverse functionality. However, there are certain proteins that remain disordered, and without any associated characteristic structure under physiological conditions, that fold specifically only while binding to another target Gunasekaran et al. (2003); Sugase et al. (2007); Wright & Dyson (1999); Yi et al. (2007). Nevertheless, most proteins behave according to the function-dictated-by-structure principle.

The above discussion illustrates the close interplay between the processes of protein folding, dynamics, conformation, intermolecular interaction, and function. In this Chapter, we will explore these structure function concepts as they relate to the design of novel pharmaceutical products. For example, these structure-dynamics-function aspects outlined above dictate action mechanisms of protein drugs (also called therapeutic protein, protein pharmaceutical, protein biopharmaceutical, or just biopharmaceutical) including their appropriate design and development to treat disease. Second, technologies like x-ray crystallography have been more difficult to apply to membrane proteins, likely the most important targets for small molecule drugs, and where information on the structural consequence of ligand binding are critical to drug development. This provides a strong interest in the development and application of reliable, sophisticated analytical techniques for thorough structural examination of therapeutic and membrane proteins in order to ensure and/or understand appropriate functionality and safety of the both small and large molecule drug development. This chapter introduces effective techniques that help realize this goal and demonstrates their application across monoclonal antibodies and membrane proteins. Monoclonal antibodies are designed to bind to specific protein targets in the cell blocking function of the targets; while membrane proteins perform essential processes in the cell, such as controlling the flow of information and materials between cells and mediating activities like nerve impulses and hormone action.

## 2. Significance

The function and efficacy of protein drug and biologic therapies is determined by the structure of the protein and its ability to interact with the surrounding partners. The interrogation and verification of the three dimensional conformation becomes critical in order to demonstrate the consistency of structure and function in biologics development. This necessitates the deployment of reliable, sensitive, and high-resolution techniques capable of examining higher order structure of such biomolecules in detail.

Biopharmaceutical manufacturers are required to demonstrate the consistency of the conformational complexity to the regulatory agencies. Traditional biophysical techniques used for this purpose include circular dichroism (CD), fluorescence, ultraviolet (UV), differential scanning calorimetry (DSC), isothermal titration calorimetry (ITC), analytical ultracentrifugation (AUC), and Fourier transform infrared spectroscopy (FTIR) Pain (2000).

These techniques provide information on a global state of the molecule, and typically output the average of overall information across the whole protein. These methods can in some cases successfully detect small changes between highly similar proteins, but the differential signal will not specifically identify the defect across the global readout. Some methods may reveal signal from only a limited number of residues (e.g., aromatic amino acid residues) in the protein and the absence of a such residues in the area of interest may lead to loss of information. Alternative sensitive, sophisticated tools for structure determination, nuclear magnetic resonance spectroscopy (NMR) and X-ray crystallography, allow both sensitive and specific probes of small structural changes Drenth (1999); Ramsey & Purcell (1952). NMR spectroscopy provides detailed structural information on proteins in solution, which is based upon distance constraints obtained from nuclear Overhauser effect Wuthrich (1990). However, these methods require high protein concentrations and can face significant challenges in the case of large proteins (in the case of NMR), and some proteins are not amenable to X-ray crystallography due to their inability to form crystals. In addition, these techniques tend to be fairly complex, and the conformation of the protein observed in the crystal is just a particular (although high resolution) conformation pulsed by the crystal lattice. This provides a strong interest to employ alternative practical methods to detect small, local differences within proteins in solution for both large and small proteins and at a range of protein concentrations. Mass spectrometry (MS) has established itself as a crucial technique in the biochemist's repository of tools over the past two decades, and many different flavors of protein MS are available with a variety of choices for sample preparation, molecular ionization, detection, and instrumentation. Structural proteomics techniques such as covalent labeling Maleknia et al. (2001); Suckau et al. (1992), hydrogen/deuterium exchange (H/DX) Wales & Engen (2006), and chemical cross-linking Back et al. (2003); when coupled with highly sensitive mass spectrometry instruments; alleviate many of the above limitations and have shown promising results in the past decade. This chapter focuses on the fundamentals of these techniques, discusses challenges and limitations experienced by each method, and concludes with the successful application examples of these analytical tools.

## 3. Techniques

The following techniques can be used to determine conformational change, binding stoichiometry, and affinity for protein-ligand interactions.

### 3.1 Hydrogen/deuterium exchange mass spectrometry(H/DX-MS)

H/DX methods were introduced in 1990s and have now powerfully established themselves for probing the biomolecular structure Bai et al. (1995); Englander & Kallenbach (1983); Hvidt & Nielsen (1966); Krishna et al. (2004); Wales & Engen (2006); Woodward et al. (1982). The principle behind the technique is that protein backbone amide hydrogens are exchangeable with deuterium atoms from the solvent surrounding the protein at specific exchange rates that can be measured experimentally. The amide hydrogens at the surface exchange very rapidly, while those buried in the core have much slower exchange rates. The backbone amide hydrogens participating in the formation of hydrogen bonds will also have relatively slower exchange rates. Hence, the rate of exchange of hydrogens provides valuable insights into the bio-molecular backbone, secondary structure and structural stability.

Fig 1 shows the overall schematics of an H/DX experiment Wales & Engen (2006). The protein under consideration is subjected to a deuterium rich environment that labels surface accessible residues, followed by quenching of the reaction. The incorporation of deuterium into the biomolecule under consideration results from the natural process of hydrogen exchange with deuterium from the surrounding environment. There are a variety of methods available for the introduction of deuterium into a peptide or protein; and various experimental strategies are used for investigating the biomolecular exchange as seen in Fig 1 . The protein can either be studied intact (for global exchange analysis), or can be digested by proteolysis (for local exchange analysis), and intact proteins or peptide fragments can be analyzed using mass spectrometry, which is able to measure the increase in mass as hydrogen atoms are exchanged for deuterium. Specific solvent accessible residues in the protein show an increased mass in the mass spectrometer readout. The experiment is repeated multiple times, each time increasing the duration of the deuterium pulse exposure to the protein, allowing for the study of deuterium exchange rate kinetics.

Although H/DX technique has shown promise for utility in biopharmaceutical studies, greater automation, seamless coupling of high performance separation, and sophisticated software for automated data interpretation is required for a more routine implementation of H/DX-MS into commercial experiments Houde et al. (2011); Wales et al. (2008). Various approaches have been developed recently towards the automated data analysis for data acquisition and post-processing Chalmers et al. (2006); Kazazic et al. (2010); Pascal et al. (2009). The H/DX-MS method has been successfully applied to study structural changes introduced by kinase activation, compare isoform-specific differences in binding to a common ligand, and to map the epitopes of monoclonal antibodies Houde et al. (2009); Lee et al. (2004); Stokasimov & Rubenstein (2009).

### 3.2 Hydroxyl-radical mediated covalent labeling mass spectrometry or protein footprinting

Another popular method for investigating macromolecular conformation in solution is protein footprinting, also called Covalent Labeling, which was invented initially to characterize the sites of DNA-protein interaction Brenowitz et al. (1986); Galas & Schmitz (1978); Humayun et al. (1977); Schmitz & Galas (1980). The technique was further extended to protein structure examination by subjecting them to limited proteolysis in conjunction with separation using SDS polyacrylamide gel electrophoresis, with the first report of protein footprinting appearing in 1988 Sheshberadaran & Payne (1988). The advent of sophisticated analytical tools such as mass spectrometry for examining cleaved fragments of proteins, significantly improved the spatial resolution of the technique. A key distinction from H/DX-MS is that most labeling reagents target side-chains, while HDX-MS specifically examines the bio-molecular backbone and protein secondary structure.

The basic principle behind hydroxyl radical mediated protein footprinting approaches for probing solvent accessible residues is similar to that of H/DX-MS technique. The overall schematic for experimental setup for a covalent labeling experiment is shown in Fig 2. The protein solution is exposed to hydroxyl radicals, generated by multiple methods, which leads to stable, covalent oxidative modifications on the surface accessible residues Hambly & Gross (2005); Maleknia et al. (2001); Sharp et al. (2003; 2004); Takamoto & Chance (2006). The chemistry of amino acid and peptide oxidation using MS revealed that in dilute aqueous solution, oxidative modification of side chains is observed in a much more predominant form as compared with backbone cleavage or cross-linking. These stable side chain modifications result in mass shifts, which can be easily revealed by isolating protein fragment and then comparing the masses to unmodified forms of the protein. Thus, labeling is followed by subjecting the protein to proteolysis and high pressure liquid chromatography coupled with mass spectrometry as in the case of H/DX-MS method. Tandem mass spectrometry (MS/MS) methods have been found to be particularly suited for further identifying and localizing the specific sites of oxidation Chance (2001); Kiselar et al. (2002); Maleknia et al. (1999). Thus, the structural resolution of covalent labeling is very high, and at the single side chain level. In the

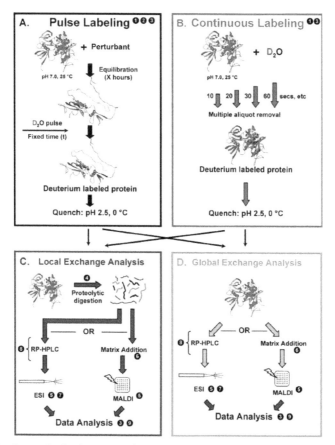

Fig. 1. Overall scheme for hydrogen exchange mass spectrometry experiments. A: Pulse labeling. After a protein has been exposed to a perturbant (chemical denaturant, heat, pH, binding, complex formation, pressure, etc.), unfolded regions (gray) become labeled with deuterium (red) during a quick pulse of D2O (typically 10 s). Deuterium exchange is quenched by reducing the pH and temperature. B: Continuous labeling. D2O buffer is added to a protein (in H2O buffer) such that the final D concentration is >95%. After a set period of time, an aliquot of the labeled protein is removed from the original tube and mixed with quench buffer to reduce the pH and temperature. Aliquot removal is repeated for subsequent labeling times. The protein concentration and solution volume are controlled such that all the aliquots are identical upon quench except for the amount of time the protein was exposed to D2O. C:. Localized exchange information. Quenched samples (from part A, part B, or both) are digested with pepsin or another acid protease. The resulting peptides are analyzed with online HPLC-ESI-MS or with MALDI-MS. The resulting data analysis provides information on deuterium exchange in short fragments of the peptide backbone. D: Global exchange information. Quenched samples (from part A, part B, or both) are directly analyzed with HPLC-ESI-MS or MALDI-MS. The data provide a global picture of how the protein behaves in D2O. Reprinted with permissions from  Wales & Engen (2006) Copyright 2006 John Wiley and Sons.

typical workflow, series of samples are exposed to variable doses, and a dose-response curve is generated for observed peptides individually in order to provide relative quantitation of oxidation as a function of hydroxyl radical exposure time. The generation of stable, covalent modifications allows a wide range of samples and proteases to be employed under broad solution conditions and pH values. The reactivity of side chains to hydroxyl radical attack initially and the attenuation of this reactivity as a result of structural perturbation such as ligand binding, unfolding, or macromolecular interactions provides insights into the change in surface accessibility at particular sites under consideration. Since the side chains get modified during the procedure, specific probe sites can be investigated using tandem mass spectrometry, while, in the case of H/DX-MS, the conformational changes may be attributed only to a specific peptide fragment. However, the two approaches are complementary to each other since H/DX-MS characterizes backbone secondary structure and stability while protein footprinting probes the side chains of residues.

The interpretation of high volumes of data resulting from covalent labeling experiments used to pose as the biggest bottleneck for the overall experiment, thus, limiting their potential. A typical hydroxyl radical-mediated covalent labeling experiment leads to multiple oxidation states of various amino acid side chains Takamoto & Chance (2006); Xu & Chance (2007), leading to a challenging task for data analysis. This bottleneck has now been eliminated with the advent of ProtMapMS, a computational analytical tool, that is specifically tailored to meet the needs of covalent labelling experiments Kaur et al. (2009). Figure 3 illustrates typical liquid chromatographic elution profile results automatedly generated in a covalent labeling experiment using ProtMapMS. The four plots in Fig 3 represent the chromatographic elution plots from a doubly charged insulin B-chain peptide 23-29 for an X-ray exposure time of 0, 8, 15, and 20 ms successively. The unoxidized form of the peptide is indicated by cyan (m/z = 430.22) color, while the different oxidative forms are blue, green, and red. Interestingly, five green peaks labeled A-E in Fig 3 represent the oxidatively labeled products for the peptide incorporating one oxygen atom represent five unique isomeric forms of the peptide molecule, differing in the position of the attached oxygen atom within the same peptide. Fig 3 shows that the relative intensities of the modified forms increase as the amount of X-ray exposure time to the protein increases. This behavior is expected since the protein molecules have increased opportunity to react with hydroxyl radicals. The oxidative forms of peptide are seen to elute at a slightly different time (although in close proximity) than their unoxidized counterpart.

Improvements in two specific areas will help in making the covalent labeling experiments more routine. More accurate quantitative relationship of solvent accessibility and the side chain reactivity is highly desirable to add quantitative rigor to the structural characterization. This will provide specific constraints that can be used in computational modeling approaches for a more comprehensive analysis. Flexible computational modeling approaches should be developed that allow for including surface accessibility constraints in a quantitative manner for more accurate results. Such improvements will lead the way for oxidative footprinting and other covalent labeling approaches using MS to be utilized for understanding the conformation and dynamics of very complex macromolecular assemblies. Footprinting technique has been very successfully applied in the past for RNA structure analysis Sclavi et al. (1998). There is a great potential for similar progress for protein structure prediction by incorporating the technique into a wider utilization.

### 3.3 Chemical cross-linking

Covalent cross-linking is another important technique for characterizing the connectivity of solution-phase complexes, and for obtaining new intramolecular or intermolecular distance constraints between biomolecules Sinz (2003); Vasilescu et al. (2004); Wine et al. (2002).

**Exposure to hydroxyl radicals and subsequent proteolysis**

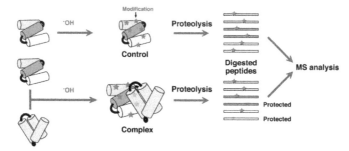

**HPLC chromatogram, ion chromatogram, and mass spectrum**

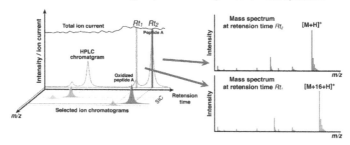

**Calculation of rate constants from dose response curve**

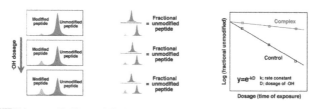

Takamoto K, Chance MR. 2006.
Annu. Rev. Biophys. Biomol. Struct. 35:251–76

Fig. 2. Hydroxyl radical footprinting: data collection and data analysis. Top panel: Protein is exposed to hydroxyl radical and modified covalently. The resulting protein sample is then digested by protease or chemical cleavage to fragments that are suitable in size for mass spectrometry. The experiment is carried out for each individual protein and for the protein complex. In a tight binding interface, some regions are protected from hydroxyl radical attack. Middle panel: Peptides are separated by liquid chromatography and introduced into a mass analyzer. The selected ion chromatograms (SIC) are constructed for each ion (with particular mass) as a function of retention time. By monitoring the mass and time, we know what species appears at what retention time. By integrating peak areas in SIC, we can calculate the total indicated ion abundance. Bottom panel: The determinations of modification rates are performed by calculating the loss of intact peptide in order to maximize the interrogation of intact material. Reprinted with permissions from Takamoto & Chance (2006) Copyright 2006 Annual Reviews.

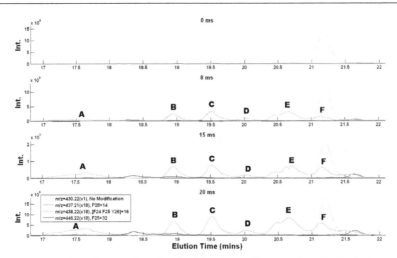

Fig. 3. Chromatographic elution plots for doubly charged human insulin B-chain peptide 23-29. The unmodified form is shown in cyan, while the modified forms (magnified by a factor of 18) are shown in blue (P28 + 14), green (mixture of F24 + 16, F25 + 16, and Y26 + 16), and red (F25 + 32). Reprinted with permission from Kaur et al. (2009). Copyright 2009 American Chemical Society.

It involves covalently attaching two specific functional groups of the protein(s) under investigation by means of a special reagent called cross-linker. The location and the identity of the created cross-links imposes a distance constraint on the location of the respective side chain sites and provides important clues on the three-dimensional conformation of the protein or a protein complex. Coupling chemical cross-linking with sensitive mass spectrometric analysis allows to characterize the position(s) of the introduced cross links for generating distance constraints. The wide variety of crosslinking reagents allow for varied specificities towards numerous functional groups such as primary amines, sulfhydryls, or carboxylic acids; and the wide range of spacer lengths offered by different cross-linking reagents allow the possibility to address a broad range of scientific questions. However, owing to the inherent complexity of the reaction mixtures, the identification of the cross-linked products can be quite tedious. The greatest challenge of utilizing chemical cross-linking and MS analysis is the lack of computational tools that can effectively interpret the enormous complexity of the reaction mixtures. There is a significant overhead of labor intensive manual processing involved in the data analysis since all the existing programs exhibit their specific limitations Sinz (2006). Some of the limiting bottlenecks have been eliminated with the advent of specialized search programs such as GPMAW, xQuest, searchXlinks, VIRTUALMSLAB, and ASAP de Koning et al. (2006); El-Shafey et al. (2006); Peri et al. (2001); Rinner et al. (2008); Wefing et al. (2006). Further progress into an integrated suite of algorithms addressing comprehensive needs for chemical cross-linking combined with mass spectrometry would greatly facilitate making it a generally applicable technique for rapid protein structure characterization for biopharmaceutical experiments.

Cross-linking and covalent labeling methods are complementary to each other - cross-linking methods provide distance constraints for two amino acids whereas covalent labeling approaches derive information about protein surface mapping since the reactions are typically controlled by the accessibility of surface amino acids to the covalent labeling reagent.

### 3.4 Future trends

Experimental data from structural proteomics experiments can be used together with computational structure modeling techniques such as comparative modeling and threading. The stand alone theoretical models without support from experimental data lack reliability, especially in the case of ab-initio modeling where suitable templates may not be available. Hybrid approaches resulting from a combination of theoretical modeling and experimental methods such as hydrogen-deuterium exchange and covalent labelling are gaining increasing popularity, allowing to combine the merits from both the methodsPantazatos et al. (2004); Zhu et al. (2003). The results from experimental analysis specifically provide explicit constraints such as distance constraints in the case of chemical cross-linking, that reflect the surface accessibility or burial of particular sites, which can be included for refining computational structure prediction models, hence greatly reducing the model space to be considered while increasing the reliability from complementary approaches.

## 4. Applications

### 4.1 Study of membrane proteins using protein footprinting

Proteins embedded in membranes assist in water or ion transport in signaling processes across the biological membrane.   Typically, transmembrane proteins are comprised of hydrophobic cores with ionizable or charged residues at specific locations that are crucial for their appropriate functionality Muller et al. (2008). G protein-coupled receptors (GPCRs) comprise a large protein family of transmembrane receptors that sense molecules outside the cell and activate the signal transduction pathways inside and regulate cellular responses Rosenbaum et al. (2009). The presence of ordered, structural waters are likely to be important factors to impart structural plasticity required for agonist-induced signal transmission for allosteric activation of the G protein-coupled receptors (GPCRs) Rosenbaum et al. (2007). The functionality of these ordered water molecules is not clearly known. They may provide structural stabilization, mediate conformational changes in signaling, neutralize charged residues, or carry out a combination of all these functions.   Structural investigation of GPCR superfamily members using radiolytic footprinting revealed the presence of conserved embedded water molecules likely to be important for GPCR function Angel, Gupta, Jastrzebska, Palczewski & Chance (2009).

The behavior of soluble proteins with hydroxyl radical footprinting is well-characterized, such that the intrinsic reactivity and the solvent accessibility of the side chains govern their observed reactivity Chance et al. (1997); Kiselar et al. (2002); Takamoto & Chance (2006).   However, these approaches have not been investigated for membrane proteins, factors influencing labeling or the overall scavenging effects of detergents or lipids have not been well understood.   Recently, in order to gain insights into membrane proteins, radiolytic protein footprinting was used to interrogate the structural dynamics of ground state (rhodopsin), photoactivated (Meta II), and inactive ligand-free receptor (opsin) and native membranes Angel, Gupta, Jastrzebska, Palczewski & Chance (2009).   In contrast to the previous literature on soluble proteins, oxidative modifications were found on residues located in both solvent-accessible and solvent-inaccessible regions.   The oxidized residues within the transmembrane domain were labeled, and their reactivity was found to be varying as a function of rhodopsin activation state.   Using radiolytic hydroxyl radical labeling in conjunction with $H_2O^{18}$ solvent mixing, it was discovered that labeling within the transmembrane region is highly influenced by the tightly bound waters and that regions undergoing local conformational alterations and water reorganization experience changes in the oxidation status.

Fig 4 shows the findings of the study, indicating alterations in rates of oxidation introduced going from ground state to activated receptor, reflecting local structural changes upon the formation of both Meta II and opsin. No exchange of the structural waters was observed with the surrounding solvent in either the ground state or for the Meta II or opsin states. However, oxidative labeling of selected side chain residues within the transmembrane helices was observed and activation-induced changes in local structural constraints likely mediated by dynamics of both water and protein were revealed. This work suggests a possible general mechanism for water-dependent communication in family A GPCRs, and illustrates the role of radiolytic footprinting for characterizing the structure and dynamics of the transmembrane region, including dynamics of water in membrane proteins, and has the potential to define allosteric channels for other transmembrane signalling proteins, and ion channels.

The implications for these results are considerable for the design of drugs to target important Type A GPCRs, e.g. serotonin, adenosine, or $\beta_2$-adrenergic receptor. If this model for rhodopsin functional activation is correct, it means that effective drugs mediate specific and local changes in water/side-chain interactions within the transmembrane region and that this rearrangement mediates the correct and efficient signaling across the membrane. In addition it focuses attention on water molecules and their potential rearrangements in recent crystallographic data of Type A family GPCRs Angel, Chance & Palczewski (2009). Recent extensions of the covalent labeling methodology to ion channel structure-function studies Gupta et al. (2010), where the movement of water coupled to the rearrangement of specific side chains was specifically tracked in the mechanism of channel opening, show the emerging power of the method to reveal important structure function considerations relevant for drug development.

### 4.2 Characterizing monoclonal antibody IgG1 using hydrogen-deuterium exchange

Monoclonal antibodies (mAb) are used both in fundamental research and in clinical settings as highly specific therapeutic agents for treating an array of different diseases. Currently, recombinant immunoglobulin gamma (IgG) mAbs comprise the largest percentage of molecules in the biopharmaceutical development pipeline. This provides a strong motivation for utilizing new or improved analytical methods and tools for mAb characterization. Presently, there is very limited information available on crystal structures of entire IgGs. Moreover, such cases provide information only of a very stable structure sampled by the protein; while lacking information on the conformational dynamics or in-solution motion.

In a recent study, H/DX-MS has been used to study both global and local conformational behavior of a recombinant monoclonal IgG1 antibody to obtain detailed conformational dynamics Houde et al. (2009). It demonstrates the capabilities of H/DX-MS as a powerful analytical tool to study large protein biopharmaceuticals such as mAbs. The conformational features of an intact glycosylated IgG1 are compared against its deglycosylated form. This assists in drawing conclusions to determine how glycosylation affects the IgG1 conformation. First, deuterium exchange into the intact form of IgG1 was measured. This is useful for providing information about the overall solvent accessibility of the protein and, also indicates whether the protein is amenable to H/DX-MS experiments so that further investigation can be performed. The next step performed the analysis of exchange into isolated Fab/Fc fragments. Next, the intact protein was labeled and digested (after quenching the deuterium labeling reaction) using pepsin as a protease. Digestion protocols were performed for both glycosylated and deglycosylated versions of the IgG1, followed by analysis using mass spectrometry. Five independent experiments were performed, each containing an undeuterated sample and five different labeling times

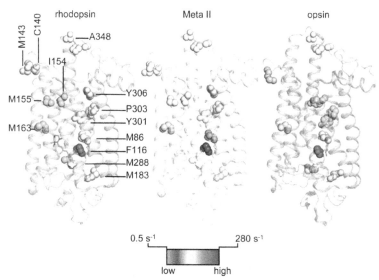

Fig. 4. Pictorial summary of modification rate constants. Radiolytic modification rate constants were determined for many residues in rhodopsin (Left), Meta II (Center), and opsin (Right). Residues with rate constants >0.1 s$^{-1}$ are rendered as spheres colored by rate constant ranges: 0.5-1.2 s$^{-1}$, light blue; 1.3-3.9 s$^{-1}$, light green; 4.0-5.9 s$^{-1}$, green; 6.0-7.9 s$^{-1}$, light-yellow; 8.0-9.9 s$^{-1}$, yellow; 10-14.9 s$^{-1}$, light-orange; 15-25 s$^{-1}$, orange; >200 s$^{-1}$, red. Following photoactivation, modification rates increased for M86, C140, M143, the pair of residues in helix IV I154 and M155, M163, and M288. Residues exhibiting decreased modification rates were Y301, P303, and Y306 in helix VII. There also was a reduced modification rate of M86 and F116 in opsin as compared with the two other states. The mixed modification of peptide 137-146, comprising part of the C-II loop, showed a large increase in the rates of detectable modification for opsin relative to ground state and activated rhodopsin, whereas M183 in the E-II loop exhibited no change in modification rate as a function of receptor activation state. The carboxyl terminal peptide did not show a marked difference in modification rates between the three states of the receptor. Changes in rates of oxidation observed when comparing ground state and activated receptor reflect local structural changes upon formation of both Meta II and opsin. Reprinted with permissions from Angel, Gupta, Jastrzebska, Palczewski & Chance (2009) Copyright 2009 National Academy of Sciences.

Figure 5 depicts the oxidative changes detected in IgG1 in the presence and absence of glycosylation. Analysis of H/D exchange pattern into the intact, glycosylated IgG1 indicated that the molecule was folded, very stable, and could be analyzed with very high sensitivity. Since the approach can detect subtle, localized changes within the protein, H/D exchange could be localized to very specific regions of the antibody. Degylosylation resulted in changes in the IgG1 conformation, and were characterized by comparing H/D exchange rates of the glycosylated and deglycosylated forms of the antibody. Two specific regions of the IgG1 (residue positions 236-253 and 292-308) were found to have experienced change in H/D exchange properties upon deglycosylation. These results are consistent with previous findings using X-Ray crystallography and NMR techniques associating the role of glycosylation in the interaction of IgG1 with Fc receptors. Overall, H/DX-MS showed that changes in

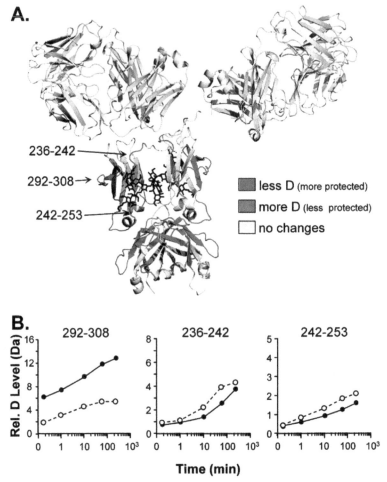

**A.**

236-242

292-308→

242-253

■ less D (more protected)
■ more D (less protected)
☐ no changes

**B.**

292-308     236-242     242-253

Rel. D Level (Da)

Time (min)

Fig. 5. Comparison of deuterium levels in IgG1 with and without glycosylation. (A) The model structure of IgG1, with the glycosylation indicated in black sticks. Parts colored blue indicate regions where the deglycosylated form had, over all time points, less deuterium (more protection from exchange). Parts colored red indicate regions where the deglycosylated form had, over all time points, more deuterium (less protection from exchange). Note that although blue regions appear to be more surface accessible than the red regions, the conformational distrubances introduced during the deglycosylation process result in greater protection in blue areas than their red counterparts (B) Representative deuterium incorporation profiles comparing exchange in heavy chain residues 292-308 (PREEQYNSTYRVVSVLT), 236-242 (LGGPSVF), and 242-253 (FLFPPKPKDTLM). The solid, black line represents data from the glycosylated form, and the dotted line represents data from the deglycosylated form. Reprinted with permission from Houde et al. (2009). Copyright 2009 American Chemical Society.

conformation as a result of deglycosylation were in areas critical for Fc receptor binding. The data illustrate the utility of H/DX-MS to provide valuable information on the higher order structure of antibodies and characterizing conformational changes that these molecules may experience upon modifications such as glycosylation, possibly affecting the functionality of the protein.

Recently, the application of covalent labeling to the analysis of glycoprotein structure was demonstrated Wang et al. (2010). This shows that covalent labeling can also be successfully applied to monoclonal antibodies to determine the structure-function relationships relevant to biopharmaceutical development.

## 5. Conclusions

There is a close inter-relation between the functionality of a protein and the processes of protein folding, dynamics, three-dimensional structure, and intermolecular interaction. These aspects further dictate action mechanisms of a protein drug and its effectiveness in the treatment of a disease. The examination of primary sequence, specific protein modifications, and the three dimensional conformation are important variables used to demonstrate structural equivalency of clinically relevant formulations for biologics development. An evolving regulatory climate is likely to require highly accurate data on "higher-order structure" of biologics for the approval of biosimilars in the US. This provides strong impetus for developing reliable, sensitive, and high-resolution analytical technologies along with associated computational methods for detailed primary and secondary structural interrogation of biomolecules. MS based techniques are proving to be indispensable tools for monitoring protein folding, structure, and dynamics. Rapid progress in this technology has allowed the possibility of highly sophisticated experiments for addressing complex biological questions.

Many different configurations of protein MS are now available with a wide range of choices for sample preparation, molecular ionization, detection, and instrumentation. This chapter introduced MS based structural proteomics techniques such as H/D exchange, oxidative covalent labeling, and chemical cross-linking. These techniques, when used in conjunction with highly sensitive mass spectrometry instruments; overcome many of the limitations experienced by traditional biophysical methods, and have gained wide popularity with promising results in the past decade. In the H/D exchange process, a covalently bonded hydrogen atom from the protein backbone is replaced by a deuterium atom from the surrounding environment, or vice versa. It provides information about the solvent accessibility of various parts of the molecule. The rate of H/D exchange imparts understanding of protein backbone and the secondary structure. Covalent labeling is based on a similar principle, except that specific labeling reagents can be used that lead to stable, covalent modifications on the solvent accessible residues. Protein footprinting is a popular covalent technique in which hydroxyl radicals are generated that create specific oxidative modifications on the surface accessible residues, which helps in mapping protein surfaces. Most covalent labeling reagents target side chains, while H/DX-MS specifically probes protein backbone and tertiary structure. Chemical cross linking involves covalently attaching two specific inter- or intra-molecular functional groups of side chains by means of a cross-linking reagent. The cross-linking agent imposes a distance constraint on the respective functional groups, providing valuable information on the three dimensional structure of a macromolecule or macromolecular assembly.

Greater automation of experimental workflow along with reliable and sophisticated computer software will pave the way for a more routine incorporation of structural mass spectrometry into commercial experiments. The future trends lie in the use of structural proteomics experimental data coupled with computational modeling techniques such as comparative modeling and threading. Hybrid approaches combining the theoretical models and experimental methods will allow to combine the merits from both approaches. Experimental results can provide specific, explicit constraints such as distance limitations or surface accessibility information, which can be included for refining computational structure prediction models for more robust and reliable results.

The chapter concludes with the discussion of successful application examples of MS based structural proteomics tools in the context of the design of novel pharmaceutical products. Structural examination of GPCR superfamily members using hydroxyl radical mediated footprinting coupled with $H_2O^{18}$ labeling showed the presence of conserved embedded water molecules that are likely to be important for GPCR function. These structural waters were not found to exchange with the surrounding solvent in either the ground state or for the Meta II or opsin states. On the other hand, oxidative modification of selected side chain residues within the transmembrane helices was detected and activation-induced changes in local structural constraints were revealed, likely to be mediated by the dynamics of both water and protein. The results suggest the possibility of a general mechanism for water-dependent communication in family A GPCRs. This example illustrates the importance and potential of radiolytic footprinting for characterizing the structure and dynamics of the transmembrane region, including dynamics of water in membrane proteins.

H/DX-MS has been used to characterize both global and local structural behavior of a recombinant monoclonal IgG1 antibody to study detailed conformational dynamics. The intact, glycosylated form of IgG1 was found to be folded and stable, while the degylosylation resulted in changes in the IgG1 conformation. This was evident by comparing H/D exchange rates of the glycosylated and deglycosylated forms of the antibody. Two specific regions of the IgG1 (residue positions 236-253 and 292-308) were found to have experienced change in exchange properties upon deglycosylation. The data illustrate the utility of H/DX-MS to provide insights into the higher order structure of antibodies and characterizing conformational changes that these molecules may experience upon modifications such as glycosylation, possibly affecting the functionality of the protein.

Covalent labeling has also shown successful deployment towards monoclonal antibodies characterization to establish the structure-function relationships in context of biopharmaceutical development. Such studies are currently ongoing at several pharmaceutical companies and are likely to have a significant impact on the development of both new drugs and biosimilars in the near future.

## 6. References

Anfinsen, C. B. (1973). Principles that govern the folding of protein chains, *Science* 181: 223–230.

Angel, T. E., Chance, M. R. & Palczewski, K. (2009). Conserved waters mediate structural and functional activation of family A (rhodopsin-like) G protein-coupled receptors, *Proc. Natl. Acad. Sci. U.S.A.* 106: 8555–8560.

Angel, T. E., Gupta, S., Jastrzebska, B., Palczewski, K. & Chance, M. R. (2009). Structural waters define a functional channel mediating activation of the GPCR, rhodopsin, *Proc. Natl. Acad. Sci. U.S.A.* 106: 14367–14372.

Back, J. W., de Jong, L., Muijsers, A. O. & de Koster, C. G. (2003). Chemical cross-linking and mass spectrometry for protein structural modeling., *J Mol Biol* 331(2): 303–313.

Bai, Y., Sosnick, T. R., Mayne, L. & Englander, S. W. (1995). Protein folding intermediates: native-state hydrogen exchange, *Science* 269: 192–197.

Brenowitz, M., Senear, D. F., Shea, M. A. & Ackers, G. K. (1986). "Footprint" titrations yield valid thermodynamic isotherms, *Proc. Natl. Acad. Sci. U.S.A.* 83: 8462–8466.

Chalmers, M. J., Busby, S. A., Pascal, B. D., He, Y., Hendrickson, C. L., Marshall, A. G. & Griffin, P. R. (2006). Probing protein ligand interactions by automated hydrogen/deuterium exchange mass spectrometry, *Anal. Chem.* 78: 1005–1014.

Chance, M. R. (2001). Unfolding of apomyoglobin examined by synchrotron footprinting, *Biochem. Biophys. Res. Commun.* 287: 614–621.

Chance, M. R., Sclavi, B., Woodson, S. A. & Brenowitz, M. (1997). Examining the conformational dynamics of macromolecules with time-resolved synchrotron x-ray 'footprinting'., *Structure* 5(7): 865–869.

de Koning, L. J., Kasper, P. T., Back, J. W., Nessen, M. A., Vanrobaeys, F., Van Beeumen, J., Gherardi, E., de Koster, C. G. & de Jong, L. (2006). Computer-assisted mass spectrometric analysis of naturally occurring and artificially introduced cross-links in proteins and protein complexes, *FEBS J.* 273: 281–291.

Dobson, C. M. (2001). The structural basis of protein folding and its links with human disease, *Philos. Trans. R. Soc. Lond., B, Biol. Sci.* 356: 133–145.

Dobson, C. M. (2003). Protein folding and misfolding, *Nature* 426: 884–890.

Drenth, J. (1999). *Principles of protein x-ray crystallography*, Springer-Verlag.

Ecroyd, H. & Carver, J. A. (2008). Unraveling the mysteries of protein folding and misfolding, *IUBMB Life* 60: 769–774.

El-Shafey, A., Tolic, N., Young, M. M., Sale, K., Smith, R. D. & Kery, V. (2006). "Zero-length" cross-linking in solid state as an approach for analysis of protein-protein interactions, *Protein Sci.* 15: 429–440.

Englander, S. W. & Kallenbach, N. R. (1983). Hydrogen exchange and structural dynamics of proteins and nucleic acids, *Q. Rev. Biophys.* 16: 521–655.

Falke, J. J. (2002). Enzymology. A moving story, *Science* 295: 1480–1481.

Fenimore, P. W., Frauenfelder, H., McMahon, B. H. & Parak, F. G. (2002). Slaving: solvent fluctuations dominate protein dynamics and functions, *Proc. Natl. Acad. Sci. U.S.A.* 99: 16047–16051.

Frauenfelder, H., Sligar, S. G. & Wolynes, P. G. (1991). The energy landscapes and motions of proteins, *Science* 254: 1598–1603.

Galas, D. J. & Schmitz, A. (1978). DNAse footprinting: a simple method for the detection of protein-DNA binding specificity, *Nucleic Acids Res.* 5: 3157–3170.

Gunasekaran, K., Tsai, C. J., Kumar, S., Zanuy, D. & Nussinov, R. (2003). Extended disordered proteins: targeting function with less scaffold, *Trends Biochem. Sci.* 28: 81–85.

Gupta, S., Bavro, V. N., D'Mello, R., Tucker, S. J., Venien-Bryan, C. & Chance, M. R. (2010). Conformational changes during the gating of a potassium channel revealed by structural mass spectrometry, *Structure* 18: 839–846.

Hambly, D. M. & Gross, M. L. (2005). Laser flash photolysis of hydrogen peroxide to oxidize protein solvent-accessible residues on the microsecond timescale, *Journal of the American Society for Mass Spectrometry* 16(12): 2057 – 2063.

Hegyi, H. & Gerstein, M. (1999). The relationship between protein structure and function: a comprehensive survey with application to the yeast genome, *J. Mol. Biol.* 288: 147–164.

Henzler-Wildman, K. & Kern, D. (2007). Dynamic personalities of proteins, *Nature* 450: 964–972.

Houde, D., Arndt, J., Domeier, W., Berkowitz, S. & Engen, J. R. (2009). Characterization of IgG1 Conformation and Conformational Dynamics by Hydrogen/Deuterium Exchange Mass Spectrometry, *Anal. Chem.* 81: 5966.

Houde, D., Berkowitz, S. A. & Engen, J. R. (2011). The utility of hydrogen/deuterium exchange mass spectrometry in biopharmaceutical comparability studies, *J Pharm Sci* 100: 2071–2086.

Huang, Y. J. & Montelione, G. T. (2005). Structural biology: proteins flex to function, *Nature* 438: 36–37.

Humayun, Z., Kleid, D. & Ptashne, M. (1977). Sites of contact between lambda operators and lambda repressor, *Nucleic Acids Res.* 4: 1595–1607.

Hvidt, A. & Nielsen, S. O. (1966). Hydrogen exchange in proteins, *Adv. Protein Chem.* 21: 287–386.

Kamerzell, T. J. & Middaugh, C. R. (2008). The complex inter-relationships between protein flexibility and stability, *J Pharm Sci* 97: 3494–3517.

Kaur, P., Kiselar, J. G. & Chance, M. R. (2009). Integrated algorithms for high-throughput examination of covalently labeled biomolecules by structural mass spectrometry, *Anal. Chem.* 81: 8141–8149.

Kazazic, S., Zhang, H. M., Schaub, T. M., Emmett, M. R., Hendrickson, C. L., Blakney, G. T. & Marshall, A. G. (2010). Automated data reduction for hydrogen/deuterium exchange experiments, enabled by high-resolution Fourier transform ion cyclotron resonance mass spectrometry, *J. Am. Soc. Mass Spectrom.* 21: 550–558.

Kiselar, J. G., Maleknia, S. D., Sullivan, M., Downard, K. M. & Chance, M. R. (2002). Hydroxyl radical probe of protein surfaces using synchrotron x-ray radiolysis and mass spectrometry., *Int J Radiat Biol* 78(2): 101–114.
URL: *http://dx.doi.org/10.1080/09553000110094805*

Krishna, M. M., Hoang, L., Lin, Y. & Englander, S. W. (2004). Hydrogen exchange methods to study protein folding, *Methods* 34: 51–64.

Lee, T., Hoofnagle, A. N., Kabuyama, Y., Stroud, J., Min, X., Goldsmith, E. J., Chen, L., Resing, K. A. & Ahn, N. G. (2004). Docking motif interactions in MAP kinases revealed by hydrogen exchange mass spectrometry, *Mol. Cell* 14: 43–55.

Maleknia, S. D., Brenowitz, M. & Chance, M. R. (1999). Millisecond radiolytic modification of peptides by synchrotron x-rays identified by mass spectrometry., *Anal Chem* 71(18): 3965–3973.

Maleknia, S. D., Ralston, C. Y., Brenowitz, M. D., Downard, K. M. & Chance, M. R. (2001). Determination of macromolecular folding and structure by synchrotron x-ray radiolysis techniques., *Anal Biochem* 289(2): 103–115.
URL: *http://dx.doi.org/10.1006/abio.2000.4910*

Moore, S. & Stein, W. H. (1973). Chemical structures of pancreatic ribonuclease and deoxyribonuclease, *Science* 180: 458–464.

Muller, D. J., Wu, N. & Palczewski, K. (2008). Vertebrate membrane proteins: structure, function, and insights from biophysical approaches, *Pharmacol. Rev.* 60: 43–78.

Pain, R. H. (2000). *Mechanisms of protein folding*, Oxford University Press.

Pantazatos, D., Kim, J. S., Klock, H. E., Stevens, R. C., Wilson, I. A., Lesley, S. A. & Woods, V. L. (2004). Rapid refinement of crystallographic protein construct definition employing enhanced hydrogen/deuterium exchange MS, *Proceedings of the National Academy of Sciences of the United States of America* 101(3): 751–756.
URL: *http://www.pnas.org/content/101/3/751.abstract*

Pascal, B. D., Chalmers, M. J., Busby, S. A. & Griffin, P. R. (2009). HD desktop: an integrated platform for the analysis and visualization of H/D exchange data, *J. Am. Soc. Mass Spectrom.* 20: 601–610.

Peri, S., Steen, H. & Pandey, A. (2001). GPMAW–a software tool for analyzing proteins and peptides, *Trends Biochem. Sci.* 26: 687–689.

Ramsey, N. F. & Purcell, E. M. (1952). Interactions between nuclear spins in molecules, *Physical Review* 85(1): 143–144.

Rinner, O., Seebacher, J., Walzthoeni, T., Mueller, L. N., Beck, M., Schmidt, A., Mueller, M. & Aebersold, R. (2008). Identification of cross-linked peptides from large sequence databases, *Nat. Methods* 5: 315–318.

Rosenbaum, D. M., Cherezov, V., Hanson, M. A., Rasmussen, S. G., Thian, F. S., Kobilka, T. S., Choi, H. J., Yao, X. J., Weis, W. I., Stevens, R. C. & Kobilka, B. K. (2007). GPCR engineering yields high-resolution structural insights into beta2-adrenergic receptor function, *Science* 318: 1266–1273.

Rosenbaum, D. M., Rasmussen, S. G. & Kobilka, B. K. (2009). The structure and function of G-protein-coupled receptors, *Nature* 459: 356–363.

Sadowski, M. I. & Jones, D. T. (2009). The sequence-structure relationship and protein function prediction, *Curr. Opin. Struct. Biol.* 19: 357–362.

Schmitz, A. & Galas, D. J. (1980). Sequence-specific interactions of the tight-binding I12-X86 lac repressor with non-operator DNA, *Nucleic Acids Res.* 8: 487–506.

Schulz, G. E. (1992). Induced-fit movements in adenylate kinases, *Faraday Discuss.* pp. 85–93.

Sclavi, B., Sullivan, M., Chance, M. R., Brenowitz, M. & Woodson, S. A. (1998). RNA Folding at Millisecond Intervals by Synchrotron Hydroxyl Radical Footprinting, *Science* 279(5358): 1940–1943.
URL: *http://www.sciencemag.org/cgi/content/abstract/279/5358/1940*

Sharp, J. S., Becker, J. M. & Hettich, R. L. (2003). Protein surface mapping by chemical oxidation: structural analysis by mass spectrometry., *Anal Biochem* 313(2): 216–225.

Sharp, J. S., Becker, J. M. & Hettich, R. L. (2004). Analysis of protein solvent accessible surfaces by photochemical oxidation and mass spectrometry., *Anal Chem* 76(3): 672–683.
URL: *http://dx.doi.org/10.1021/ac0302004*

Sheshberadaran, H. & Payne, L. G. (1988). Protein antigen-monoclonal antibody contact sites investigated by limited proteolysis of monoclonal antibody-bound antigen: protein "footprinting"., *Proc Natl Acad Sci USA* 85(1): 1–5.

Sinz, A. (2003). Chemical cross-linking and mass spectrometry for mapping three-dimensional structures of proteins and protein complexes, *J Mass Spectrom* 38: 1225–1237.

Sinz, A. (2006). Chemical cross-linking and mass spectrometry to map three-dimensional protein structures and protein-protein interactions, *Mass Spectrom Rev* 25: 663–682.

Stokasimov, E. & Rubenstein, P. A. (2009). Actin isoform-specific conformational differences observed with hydrogen/deuterium exchange and mass spectrometry, *J. Biol. Chem.* 284: 25421–25430.

Suckau, D., Mak, M. & Przybylski, M. (1992). Protein surface topology-probing by selective chemical modification and mass spectrometric peptide mapping., *Proc Natl Acad Sci USA* 89(12): 5630–5634.

Sugase, K., Dyson, H. J. & Wright, P. E. (2007). Mechanism of coupled folding and binding of an intrinsically disordered protein, *Nature* 447: 1021–1025.

Takamoto, K. & Chance, M. R. (2006). Radiolytic protein footprinting with mass spectrometry to probe the structure of macromolecular complexes., *Annu Rev Biophys Biomol Struct* 35: 251–276.
URL: *http://dx.doi.org/10.1146/annurev.biophys.35.040405.102050*

Teilum, K., Olsen, J. G. & Kragelund, B. B. (2009). Functional aspects of protein flexibility, *Cell. Mol. Life Sci.* 66: 2231–2247.

Travaglini-Allocatelli, C., Ivarsson, Y., Jemth, P. & Gianni, S. (2009). Folding and stability of globular proteins and implications for function, *Curr. Opin. Struct. Biol.* 19: 3–7.

van der Kamp, M. W., Schaeffer, R. D., Jonsson, A. L., Scouras, A. D., Simms, A. M., Toofanny, R. D., Benson, N. C., Anderson, P. C., Merkley, E. D., Rysavy, S., Bromley, D., Beck, D. A. & Daggett, V. (2010). Dynameomics: a comprehensive database of protein dynamics, *Structure* 18: 423–435.

Vasilescu, J., Guo, X. & Kast, J. (2004). Identification of protein-protein interactions using in vivo cross-linking and mass spectrometry, *Proteomics* 4: 3845–3854.

Wales, T. E. & Engen, J. R. (2006). Hydrogen exchange mass spectrometry for the analysis of protein dynamics, *Mass Spectrom. Rev.* 25: 158–70.

Wales, T. E., Fadgen, K. E., Gerhardt, G. C. & Engen, J. R. (2008). High-speed and high-resolution UPLC separation at zero degrees Celsius, *Anal. Chem.* 80: 6815–6820.

Wang, L., Qin, Y., Ilchenko, S., Bohon, J., Shi, W., Cho, M. W., Takamoto, K. & Chance, M. R. (2010). Structural analysis of a highly glycosylated and unliganded gp120-based antigen using mass spectrometry, *Biochemistry* 49: 9032–9045.

Wefing, S., Schnaible, V. & Hoffmann, D. (2006). SearchXLinks. A program for the identification of disulfide bonds in proteins from mass spectra, *Anal. Chem.* 78: 1235–1241.

Wine, R. N., Dial, J. M., Tomer, K. B. & Borchers, C. H. (2002). Identification of components of protein complexes using a fluorescent photo-cross-linker and mass spectrometry, *Anal. Chem.* 74: 1939–1945.

Woodward, C., Simon, I. & Tuchsen, E. (1982). Hydrogen exchange and the dynamic structure of proteins, *Mol. Cell. Biochem.* 48: 135–160.

Wright, P. E. & Dyson, H. J. (1999). Intrinsically unstructured proteins: re-assessing the protein structure-function paradigm, *J. Mol. Biol.* 293: 321–331.

Wuthrich, K. (1990). Protein structure determination in solution by NMR spectroscopy, *The Journal of biological chemistry* 265(36): 22059–22062.
URL: *http://www.jbc.org/cgi/content/abstract/265/36/22059*

Xu, G. & Chance, M. R. (2007). Hydroxyl radical-mediated modification of proteins as probes for structural proteomics, *Chemical Reviews* 107(8): 3514–3543.
URL: *http://pubs.acs.org/doi/abs/10.1021/cr0682047*

Yi, S., Boys, B. L., Brickenden, A., Konermann, L. & Choy, W. Y. (2007). Effects of zinc binding on the structure and dynamics of the intrinsically disordered protein prothymosin alpha: evidence for metalation as an entropic switch, *Biochemistry* 46: 13120–13130.

Zhu, M. M., Rempel, D. L., Du, Z. & Gross, M. L. (2003). Quantification of protein-ligand interactions by mass spectrometry, titration, and H/D exchange: PLIMSTEX, *Journal of the American Chemical Society* 125(18): 5252–5253.
URL: *http://pubs.acs.org/doi/abs/10.1021/ja029460d*

# Part 5

## Bioinformatics Tools

# 10

# Application of Bioinformatics Tools in Gel-Based Proteomics

Kah Wai Lin, Min Jia and Serhiy Souchelnytskyi
*Department of Oncology-Pathology,*
*Karolinska Institutet, Stockholm*
*Sweden*

## 1. Introduction

Personalized medicine is the most promising approaches in the treatment of various diseases, especially cancer. The use of appropriate biomarkers for personalized treatment has advantage over conventional therapeutics approach, as it confer maximum effectiveness with minimum side effect. Personalized treatment can be achieved by implementation of omic studies in clinical practices. Application of genomic, transcriptomic, proteomic and metabolomic studies deliver a vast amount of data that lead to the discovery of novel biomarkers for diagnostic, prognostic and therapeutic purposes. Therefore, further exploration in omic study could lead to the implementation of personalized medicine as a standard therapeutic scheme in the clinic.

Proteomics is a global study of entire proteins of cell, tissue and organism in a particular condition and time point (Graves & Haystead, 2002). Proteomics is a very comprehensive discipline that includes the study of expression, function, localization, structure, modification, and protein-protein interaction (Graves & Haystead, 2002; Lim & Elenitoba-Johnson, 2004). A proteomics experiment generates vast amount of data that require further analysis, and systems biology is the main approach. Systems biology is an integrative science that studies the complex behavior of biological entities at the systems level (Kitano, 2002a, 2002b). Integrating the proteomics data into systems biology language is an important approach in understanding the behavior of the complex organisms at various levels (Souchelnytskyi, 2005). In recent years, our knowledge of proteomics and system biology is growing rapidly and create an excitement in scientific community because of its potential in novel biomarker and drug discovery (Duncan & Hunsucker, 2005).

Proteomics studies are highly dependent on the technology for protein separation and identification, and bioinformatics for data analysis. By protein separation techniques, gel-based and liquid chromatography (LC)-based approaches represent the primary stream in proteomics. In gel-based approach, that is, conventional 2D gel electrophoresis (2D-GE) and 2D differential gel electrophoresis (2D-DIGE), the proteins are separated by their molecular weight and isoelectric point. In LC-based approach, the proteins or peptides are separated by using high performance liquid chromatography (Aebersold & Mann, 2003; Cravatt et al., 2007). The identification and characterization of proteins or peptides by mass spectrometry are followed after separation (Kolker et al., 2006). In more recent years, antibody-based methods emerging as important approaches in proteomics. These approaches included the use of immunohistochemistry (IHC) on tissue microarrays (TMAs), reverse phase protein arrays (RPPAs) and serum-based diagnostic assays using antibody arrays (Borrebaeck & Wingren, 2007; Brennan, O'Connor et al., 2010; Wingren & Borrebaeck, 2004).

In the present article, we focus our discussion on the various ways of translating gel-based proteomics data into systems biology using different bioinformatics approaches. Firstly, we will discuss the dataset from gel-based proteomics, including the acquisition of primary data and type of data for bioinformatics analysis. In the subsequent section, we will discuss the several way of analyzing the data acquired from gel-based proteomics, which included the ontological-based classification, hierarchical clustering, systems and network analysis (Table 1). We will focus our discussion on the general concepts of the analysis, type of

| Ontological Classification | Hierarchical Clustering | Systems and Network Analysis |
|---|---|---|
| **Query Tools**<br><br>GO-TermFinder<br>(Boyle et al., 2004)<br><br>AmiGO<br>(Carbon et al., 2009)<br><br>MatchMiner<br>(Bussey et al., 2003)<br><br>**Visualization Tools**<br><br>GoMiner<br>(Bussey et al., 2003)<br><br>FatiGO<br>(Al-Shahrour et al., 2004, 2007)<br><br>Onto-Express<br>(Draghici et al., 2003; Khatri et al., 2002)<br><br>GOSurfer<br>(Zhong et al., 2004)<br><br>GOTM<br>(Zhang et al., 2004) | Cluster+TreeView<br>(Eisen et al., 1998)<br><br>PermutMatrix<br>(Caraux & Pinloche, 2005)<br><br>POMELO II<br>(Morrissey & Diaz-Uriarte, 2009)<br><br>Genesis (Sturn et al., 2002) | Osprey<br>(Breitkreutz et al., 2002)<br><br>BioLayout<br>(Enright & Ouzounis, 2001)<br><br>CellDesigner<br>(Funahashi et al., 2007)<br><br>Cytoscape<br>(Kohl et al. 2011) |

Table 1. List of bioinformatics tools that are commonly used for gel-based proteomics.

datasets used and bioinformatics software. Some examples of studies and future directions are presented for each approach.

## 2. Dataset in gel-based proteomics

The general workflow of bioinformatics analysis of gel-based proteomics is shown in figure 1. In gel-based proteomics, various types of datasets can be generated. There can be an annotated 2D gel, mass spectra, and list of identified proteins (Taylor et al., 2003). These dataset can be qualitative or quantitative. In this review, we focus on the analysis of 2 type of datasets generated from annotated 2D gel, i.e. global expression profile and differential expression profile.

By identifying the protein spots on a 2D gel, a comprehensive, global protein expression profile can be generated. This approach can deliver a list of proteins expressed in a cell or tissue in a particular condition, which is exceptionally useful in understanding their biological characteristic. An example is a recent study on proteome profiling of breast epithelial cells with various proliferation potential. This study generate the most comprehensive 2D protein expression map with 183 proteins identified in 184A1 cells and 318 proteins identified in MCF10A cells, which lead to the understanding of their biological properties and delivered a list of potential biomarkers of early event of tumorigenesis (Bhaskaran et al., 2009).

By identifying the protein spots in 2D gels that are different in their staining intensity in different conditions, a differential expression profile can be generated. Various biological questions can be addressed by differential expression analysis. The proteome changes upon drugs treatment can be studied by comparing the 2D gel of a particular cell treated with and without drugs. For example, cellular response to histone deacetylase inhibitor in colon cancer cells was evaluated by such approach (Milli et al., 2008). Besides, various disease stages can also be compared, for example, a list of proteins were identified to be differentially regulated between normal liver tissue and hepatocellular carcinoma (Corona et al. 2010). Furthermore, the dynamic changes of proteome can also be studied. By comparing the differential expressed proteins in the neuroblastoma grown in mice in different time interval reveal the proteome changes of the disease progression and effect of host-tumor interaction (Turner et al., 2009). Therefore, differential expression analysis of 2D gels often called comparative proteomics.

By applying various systems biology analysis tools, these proteomics dataset can further improve our insight into particular biological questions. The first objective of gel-based proteomics data mining is to search for protein of biological importance, such as diagnostic biomarker and potential drug target. By comparing two or more predefined biological conditions, we can precisely define the proteins of interest among thousands of spots in the 2D gel (Meunier et al., 2007). This can be achieved by using differential expression proteome profile, or by comparative analysis of two or more global protein expression profiles. The second objective of gel-based proteomics data mining is to use clustering approach to group or classify the proteins. This is important for understanding the complex biological systems, such as classification of tumor according to the expression of proteins, for the diagnostics and therapeutics purposes (Meunier et al., 2007). This approach can be achieved by applying the bioinformatics tools on both differential expression and global expression profile. In the subsequent section, we will discuss the analysis of gel-based proteomics dataset by using various approaches, and their biological significance.

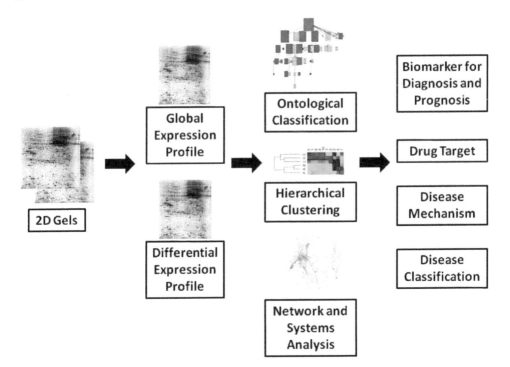

Fig. 1. General workflow of bioinformatics analysis of gel-based proteomics. Once the 2D gels are generated, 2 type of dataset can be acquired from annotated gel, i.e. global expression and differential expression profiles. These datasets can be used for further analysis by various approaches, such as ontological classification, hierarchical clustering and systems/network analysis. These analysis approaches can improve our insight into particular biological questions, such as discovery of novel disease biomarkers for diagnosis and prognosis, drug target, study of disease mechanism and disease classification.

## 3. Ontological classification

The postgenomic era has brought an exponential growth of biological databases. In recent years, researchers have begun to use unique identifiers to describe components of a database, and the relationship between them. The concept of unique identifiers forms the basis of ontology. Ontology can be described by a set of representative, unambiguous and non-redundant vocabulary or identifier, which define classes, relations, functions, objects and theories (Gruber, 1993). It is not only represents an individual component but also its related components. For instance, in anatomy ontology, stomach is define as an organ with cavity which continuous proximally with oesophagus and distally with small intestine; it is member of viscera of abdomen; it is part of gastrointestinal tract; it is supplied by left and right gastric artery; etc (Detwiler et al., 2003).

The Open Biomedical Ontology (OBO) consortium (http://www.obofoundry.org/) provides a resource where biomedical ontologies are presented in a standard format. Ontology-based approaches for data integration provide a platform of communication between researchers. It also allowed the retrieval/query of information across multiple resources and more efficient data mining and exploration. To gain the functional insight in a large-scale proteomics study, the traditional "literature mining" method is laborious and inefficient. Therefore, ontology-based approach is an effective solution.

In gel-based proteomics, the large dataset can be annotated and explored by application of Gene Ontology (GO) (http://www.geneontology.org/). Gene Ontology is a part of the Open Biomedical Ontologies (OBO), which is the most widely used ontology in biomedical research community (Smith et al., 2007). The main objective of GO is to produce a controlled and unified vocabulary for genes and gene products, such as proteins, that can be applied to all organisms. Furthermore, classification of these components in defined groups or classes allowed us to gain the functional insight in the large-scale proteomics data.

GO annotation organizes genes or gene products into hierarchical order based on 3 categories: cellular component, biological process and molecular function (The Gene Ontology Consortium, 2000). Cellular component describe the localization of particular active gene products in the cells or its extracellular environment. It may be particular cellular structure, e.g. mitochondrion, Golgi apparatus; or gene products groups, e.g. proteosome, ribosome. Biological process describes the biochemical reaction of gene products in the cells. Examples of higher order categories are cell death, signal transduction. Examples of lower order categories are lipid metabolism, purine metabolism. Molecular function describes the elemental activities of gene products at molecular levels. Examples of higher order categories are enzyme, cytoskeletal regulator. Examples of lower order categories are glycine dehydrogenase, apoptosis activator. Since March 2007, 25,000 unique GO identifiers have been created, these provide researchers a broad set of descriptors for cellular component, biological process and molecular function for genes and their products (Dimmer et al., 2008).

There are various GO tools available (table 1). The complete list of tools can be found in http://www.geneontology.org/. These tools belong to either query tools or visualization tools. Prior to analysis, the genes or proteins have to be converted from generic or common name into the unique identifier, i.e. GO term, by using query tools. The most commonly used query tools are GO-TermFinder (Boyle et al., 2004), AmiGO (Carbon et al., 2009), and MatchMiner (Bussey et al., 2003). For example, the GO identifier for cyclin D3 is CCND3.

Once the list of GO identifiers are generated, visualization the data are carried out, using the tools such as GoMiner (Bussey et al., 2003), FatiGO (Al-Shahrour et al., 2004, 2007), Onto-Express (Draghici et al., 2003; Khatri et al., 2002), GOSurfer (Zhong et al., 2004), and GOTM (Zhang et al., 2004). These tools provide visualization of data in the form of either AmiGo view or Direct Acyclic Graph (DAG) view (Figure 2). AmiGO view is in the form of expandable tree structures, and it is linked to external databases, such as NCBI and CGAP. DAG is similar to hierarchies but differ in that a more specialized and narrower term or "child" can be related to more than one less specialized and broader term or "parent". Each term are represented by a node and they connected by path in hierarchical order. Each node can often be reached from multiple paths, which allow the comparison of genes/gene products involved in more than one molecular function or biological processes.

In gel-based proteomics, data generated from global expression and differential expression profiles can be used for ontological-based classification. Many studies suggested that ontological classification is a powerful tool in functional characterization of the cells in gel-based proteomics studies. For instance, a study from Alfonso et al. showed the use of

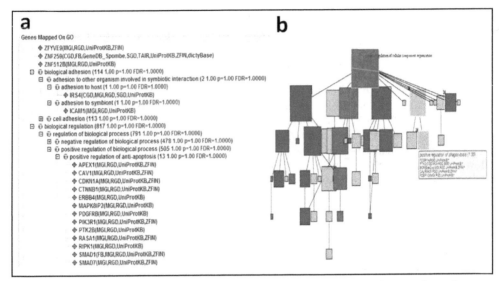

Fig. 2. Data visualization of ontological-based classification. Gene Ontology tools, such as GoMiner (Bussey et al., 2003) showed in this figure, provide visualization of data in the form of either AmiGo view or Direct Acyclic Graph (DAG) view. (a) AmiGO view is in the form of expandable tree structures, and it is linked to external databases. (b) In DAG view, each GO term are represented by a node and they connected by path in hierarchical order. Each node can often be reached from multiple paths, which allow the comparison of genes/gene products involved in more than one category.

ontological classification in a gel-based proteomics study to provide a functional insight of the colorectal cancer. In this study, 41 out of 52 analyzed proteins were unambiguously identified as being differentially expressed in colorectal cancer (Alfonso et al., 2005). An ontology analysis of these proteins revealed that they were mainly involved in regulation of transcription, cellular reorganization and cytoskeleton, cell communication and signal transduction, and protein synthesis and folding (Alfonso et al., 2005). Another example is the study of proteome changes in human T cells during peak HIV infection using 2D differential gel electrophoresis. In this study, ontological classification showed that very high proportion of differentially expressed mitochondrial and metabolic pathway proteins were identified, suggesting that metabolic reprogramming occurs upon HIV infection of T cells (Ringrose et al., 2008).

Although current proteomics study benefit from using Gene Ontology, the major drawback is that Gene Ontology does not describe and annotate the multiple forms of a gene, such as alternative slicing, proteolytic cleavage and post-translational modification. Therefore, Gene Ontology cannot describe the functional stage of the gene products. In recent year, Protein Ontology (PRO) database has been created, which provide a formal classification of proteins (Natale et al., 2007, 2011; Reeves et al., 2008). The PRO included the classification of proteins based on the basis of evolutionary relationships and the structured representation of multiple protein forms of a gene. An initial attempt in applying PRO for the annotation of TGF-beta signalling proteins showed that PRO provide a more accurate annotation and also facilitate various analysis, such as cross-species analysis, pathway analysis and disease modelling (Arighi et al., 2009). Despite of that, implementation of PRO in proteomics study is still in the infancy stage and there is no tools developed for the analysis of large-scale proteomics data. This implicates that further refinement and development of tools for PRO is needed in order to fill the gap.

## 4. Hierarchical clustering

Hierarchical clustering is a powerful approach for analyzing and visualizing the large proteomics dataset. Cluster analysis was initially designed for transcriptomics studies, such as analysis of microarray data, to explore the similarity between samples based on the pattern of gene expression (Eisen et al., 1998). In recent years, the hierarchical clustering has been adapted to the proteomics study. It enables the proteins to be grouped or classified blindly according to their expression profiles. It is a useful approach in understanding the interdependencies of protein in expression profile, molecular classification and protein signature discovery of diseases, and the dynamic changes of protein expression.

The major principle of hierarchical clustering is based on the dissimilarity or distance between the samples. In proteomics data analysis, this can be calculated by using Pearson correlation coefficient or Euclidean distance. Once the distant matrix is calculated, agglomerative clustering algorithm is performed. In proteomics, unweighted paired group average linkage (UPGMA), complete linkage, and Ward's methods are the most commonly used algorithms. The final results are presented as dendrogram or heat map (Meunier et al., 2007).

In dendrogram, proteins which are closely related will appear on the same branches. The length of branch represents the strength of relationship, where shorter the branch, closer the relationship. In a heat map, group of similar expression will appear as a pattern of cluster with same color. In either presentation method, the ultimate aim is to find the cluster which indicates a similar biological function related to disease mechanism for diagnosis and prognosis purpose.

There are several tools available for hierarchical clustering, for example, Cluster+TreeView (Eisen et al., 1998), PermutMatrix (Caraux & Pinloche, 2005), POMELO II (Morrissey & Diaz-Uriarte, 2009), and Genesis (Sturn et al., 2002). However, most currently available tools are mainly developed for transcriptomics study, i.e. analysis of cDNA microarray data. They are based on different algorithms, and only some of them can be well adapted to the proteomics data analysis, such as Cluster+TreeView and PermutMatrix (Eisen et al., 1998).

The general workflow of hierarchical clustering analysis using PermutMatrix, is discussed here. The proteomics data is presented in the form of standard text file that contains the data matrix: columns represent the sample, i.e. gels with various biological classes or groups, and row represent proteins of interests. Thereafter, the selection of clustering parameters for both distance and aggregation procedures, followed by the application of hierarchical clustering analysis. The result of clustering can be visualized in the form of dendrogram of gel samples and proteins, and heatmap of the clustered data matrix (Meunier et al., 2007) (Figure 3).

Many studies proven that hierarchical clustering is a powerful tool in analysis of large proteomics dataset. Hierarchical clustering can be use for analysis of differential expression protein or global protein expression profiles from the 2D gel. Studies suggested that hierarchical clustering is a powerful tool for discovery of protein signatures or cluster of proteins for molecular classification of diseases, especially cancer. These was shown in

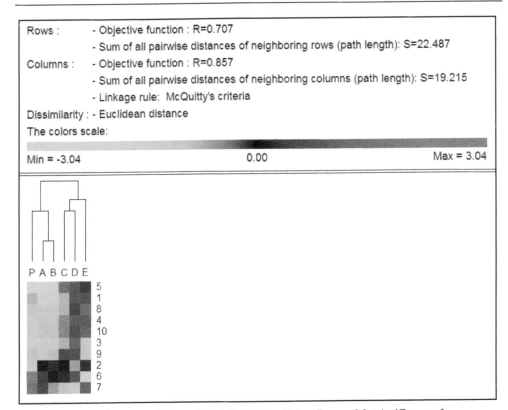

Rows :          - Objective function : R=0.707
                - Sum of all pairwise distances of neighboring rows (path length): S=22.487
Columns :       - Objective function : R=0.857
                - Sum of all pairwise distances of neighboring columns (path length): S=19.215
                - Linkage rule: McQuitty's criteria
Dissimilarity : - Euclidean distance
The colors scale:

Min = -3.04                                    0.00                                    Max = 3.04

Fig. 3. Data visualization of hierarchical clustering. Using PermutMatrix (Caraux & Pinloche, 2005), hierarchical clustering are presented as dendrogram or heat map. In dendrogram, proteins which are closely related will appear on the same branches. The length of branch represents the strength of relationship, where shorter the branch, closer the relationship. In heat map, group of similar expression will appear as a pattern of cluster with same color.

several recent studies that hierarchical clustering facilitates accurate molecular classification of vaginal and cervical cancer (Hellman et al., 2004), ovarian cancer (Alaiya et al., 2002), lung cancer (Wingren & Borrebaeck, 2004), soft-tissue sarcoma (Suehara et al., 2006), based on their protein expression profile from 2D gel. These studies might lead to the discovery of tumour-specific markers among the differentially expressed proteins. Besides, hierarchical clustering facilitates the discovery of protein signature for prediction of disease progression. This was shown by the study of a set of 20 protein spots that could predict the survival of patients with lung adenocarcinoma (Chen et al., 2003).

Many studies showed the similarity in methodology between transcriptomics and proteomics data analysis using hierarchical clustering approach. As such, many bioinformatics tools that are developed for microarray study can be adapted to gel-based proteomics studies. However, special attentions are needed, as not all the algorithms used for transcriptomics study can be used in proteomics study (Meunier et al., 2007). Without strong knowledge of these algorithms, hierarchical clustering analysis of proteomics data could lead to false result and ambiguity. This implicate that the development of new tools of hierarchical clustering analysis for proteomics study is needed to fulfil the demand of ever-growing proteomics society.

## 5. Systems and network analysis

The behaviour of a biological system, such as cells, is the consequence of complex interaction between their individual components, such as DNAs, proteins, metabolites, and other biological active molecules. In the past decades, signalling pathway has been the only approach to understand the interaction between these components. However, it is impossible to predict the behaviour of biological systems solely from understanding of their individual component or single signalling pathway. Integration of signalling pathways into a higher order biological network is a very crucial approach for studying the complex behaviour of a biological system. These can be achieved by implementation of systems and network analysis tools. In addition, the recent success of genomics and proteomics technologies generates a vast amount of data that has increased the quest for the systems and network analysis tools.

Over the past few years, application of system and network analysis in genomics and proteomics study had showed a great promise in understanding of complex behaviours of biological systems. Global mapping of the cells or organelles using these tools enable us to discover, visualize and explore the behaviour of the biological systems relevant to our experimental design. In addition, by studying the topological, functional, and dynamic properties of biological networks, the regulatory and control mechanism of the cells underlying the changes of environment can be explored. An example is a study of the over-expression of certain signalling pathway of the tumor cells under the challenge with chemotherapeutics drug (Barabasi & Oltvai, 2004; Kwoh & Ng, 2007).

Networks are displayed as graphs, which represented by nodes and edges/links. These graphs differ from the ontological and hierarchical clustering in that each node is not a function, but a component, such as gene or protein; or a substrate/product of a reaction. Nodes are displayed in various shapes, which represent various types of molecules, such as genes, proteins, and metabolites. The nodes are connected with each other by the edges or links. Edges or links represent the biological relationships between the nodes, such as induction, activation, inhibition, post-translational modification, enzymatic-substrate reaction, and physical binding.

The interaction between the nodes can be directed or undirected. In directed network, the link between two nodes has a defined direction, for example, the induction of activation of a protein by an enzyme. In undirected network, the link does not have specific direction, for example, protein-protein interaction or physical binding. Network can provide a framework from which complex regulatory information can be extracted. Most of the biological networks are scale-free, in which most of the nodes have only a few links, while a few nodes with a very large number of links, which are called hubs (Barabasi & Oltvai, 2004).

The general principle of network construction is based on the known interaction pair of gene or protein. In brief, Swiss-Prot and GeneBank accession numbers from the experimental dataset are used to search against the external databases that contain information about the interaction between the genes or proteins. Subsequently, the genes or proteins from the experiment data were integrated and merged with their known interacting partners and pathways. This process is continued until all proteins of interest are included into the network.

There are a number of available tools for construction and analysis of networks (Thomas & Bonchev, 2010), such as Osprey, (Breitkreutz et al., 2002), BioLayout (Enright & Ouzounis, 2001), CellDesigner (Funahashi et al., 2007), and Cytoscape (Kohl et al. 2011; Smoot et al. 2011). Each tool has distinct functional features. Although most of these tools were initially designed for genomics data analysis, most of them are well adapted for proteomics data analysis. For gel-based proteomics, both global expression profile and differential expression profile can be used to construct the network, depending on the experimental design and question to be answered.

Here we show an example of workflow of network analysis in gel-based proteomics, by using Cytoscape. Cytoscape is open source software that provides basic functionality for integrating proteomics data on the network, editing and visualization of network, and also implementation of external plug-ins for network analysis. Data generated from gel-based proteomics, i.e. the list of the proteins, are integrated with the graph using tools for network construction, such as MiMi (Gao et al., 2009), cPath (Cerami et al., 2006) and BioNetBuilder (Avila-Campillo et al., 2007). Subsequently, using the annotation tools, the node and edge can be annotated with attribute and expression data, such as expression ratio obtained from 2D gel analysis. For visualization of network structure, Cytoscape supports a variety of network layout algorithms, such as spring-embedded layout, circular layout and hierarchical layout (Figure 4).

In order to reduce the complexity of a large network, user can selectively display the set of nodes and edges in the graph, using graph selection and filtering tools. Nodes and edges can be selected according to a wide variety of criteria, including selection by name or by the property of the attribute (Figure 5). Besides, Cytoscape are supported by filtering tools that includes a Minimum Neighbors filter, Local Distance filter, Differential Expression filter, or the combination filter. Minimum Neighbors filter selects nodes having a minimum number of neighbors within a specified distance in the network. Local Distance filter selects nodes within a specified distance of a group of nodes. Differential Expression filter selects nodes according to their expression data. Combination filter selects nodes by combinations of other filters (Shannon et al., 2003).

When the network construction is complete, user can implement various external plug-ins for analysis of the network. This is one of the most powerful functionality of Cytoscape for solving biological questions by mean of network exploration. There is a variety of plug-ins which is commonly used in network analysis. Several examples of Cytoscape plug-ins for network analysis, such as MCODE (Bader & Hogue, 2003), NetworkAnalyzer (Assenov et al., 2008) and Centiscape (Scardoni et al., 2009), are discussed here. MCODE is a plug-in that search for clusters or highly interconnected regions in the network (Bader & Hogue, 2003). In protein network, clusters are often attribute to a groups of proteins that represent a proteins family or protein-protein interaction networks, therefore, finding the cluster enable us to define the region of functional importance. NetworkAnalyzer is a Java plug-in that analyses and visualizes the molecular interaction networks (Assenov et al., 2008). NetworkAnalyzer computes different parameters that describe the network topology, such as diameter of a network, average number of neighbours, and numbers of connected pairs of nodes. NetworkAnalyzer also compute more complex parameters, for example, node degree distribution, topological coefficients, shortest path length distribution, closeness centrality and neighbourhood connectivity distribution. These topology parameters enable us to understanding the property of biological network, such as protein signalling network, protein-protein interaction network, that are of biological importance. Centiscape is another

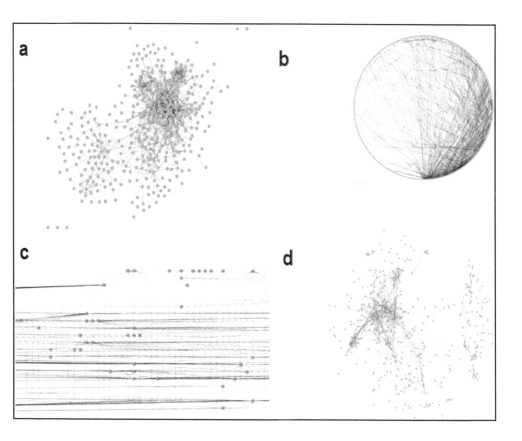

Fig. 4. Visualization of network structure using Cytoscape (Kohl et al. 2011; Smoot et al. 2011). Networks are displayed as graphs, which represented by nodes and edges. For visualization of network structure, Cytoscape supports a variety of network layout algorithms, such as (a) force-directed layout, (b) circular layout, (c) hierarchical layout, and (d) spring-embedded layout.

plug-in for analysis of complex topology of biological network (Scardoni et al., 2009). Centiscape computes centrality indexes of each node in the network, and relationship between the nodes. Thus, Centiscape provides classification of nodes according to their capability to influence the function of other nodes within the network. This may enable us to identify the critical nodes and regulatory circuits in the protein network.

In gel-based proteomics, network construction and pathway analysis are very useful in identifying novel regulatory mechanism of diseases and drug target discovery (Dudley & Butte, 2009). This was showed by a recent study that network analysis of proteomics data from clear cell renal cell carcinoma patient revealed the role of TNFα in clear cell renal cell carcinoma pathogenesis. In addition, it was suggested that clinically available TNFα

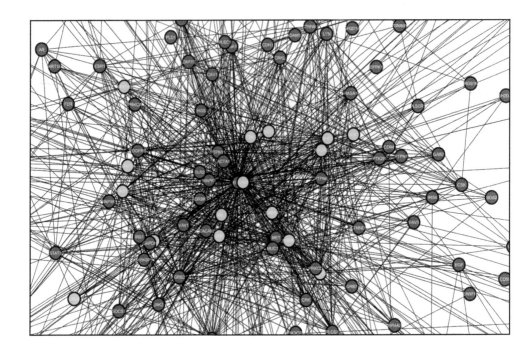

Fig. 5. Graph selection tool in Cytoscape (Kohl et al. 2011; Smoot et al. 2011). User can use graph selection tool to reduce the complexity of the graph. In this example, the components of ERBB pathway were selected and coloured (green) using selection tools.

inhibitors, such as thalidomide and etanercept can be used for the treatment of renal cell carcinoma (Perroud et al., 2006). Besides, network analysis is an indispensable tool in understanding the complex biological behaviour of the cells. A recent study showed that network analysis of gel-based proteome reveal the similarities in regulatory mechanism by MCF10A and 184A1 cells. Network analysis showed the involvement of TNF, AKT, F2 and IGF hubs in both cell types, but cell cycle regulation and mitogenic signaling networks are more representative in MCF10A cells, as compared to 184A1 cells. Study of the network also showed that enhanced expression of cell cycle and proliferation-related proteins, such as CDK4 and cyclin D3 may have an important contribution to increased proliferation rate of breast epithelial cells at the early event of tumorigenesis (Bhaskaran et al., 2009).

Network and pathway analysis is a robust approach in analyzing large proteomics dataset. However, there are several major limitations. Network analysis is unbiased and hypothesis-free because the built of network are based on known interaction sets that recruited from published data. As a consequent, network analysis is not able to uncover the new or unknown pathway and interaction. On the other hand, the qualities of network are dependent on the limitation of high-throughput experiments where the data were recruited from. For instance, protein-protein interaction studies that generate a high proportion of false-positive result will affect the quality of network based on this data (Arrell & Terzic, 2010). Nevertheless, network analysis remains a powerful tool in understanding the gel-based proteomics data, and it can serve as a good starting point for a further exploration of the dataset.

## 6. Concluding remarks

Tremendous effort have been made during past decade in understanding the biology of normal and diseased cells at systemic level. Proteomics is one of the most promising approaches in generating functional insight of biological systems. Recent advancement in protein separation and identification technology leads to the generation of enormous amount of data which implicate that importance of bioinformatics analysis. However, this renders a great challenge for biomedical researchers in selecting the suitable strategies in bioinformatics analysis of proteomics data.

This article gives an overview of various analysing strategies in gel-based proteomics; we hope that this will help biomedical researchers to derive more biologically meaningful information from their data. These effort will render a direct impact in the in-depth understanding of biological behaviour of cells, ultimately implemented in clinical applications.

## 7. References

Aebersold, R. & Mann, M. (2003). Mass spectrometry-based proteomics. *Nature,* Vol.422, No.6928, pp. 198-207, ISSN 0028-0836

Al-Shahrour, F.; Diaz-Uriarte, R. & Dopazo, J. (2004). FatiGO: a web tool for finding significant associations of Gene Ontology terms with groups of genes. *Bioinformatics,* Vol.20, No.4, pp. 578-580, ISSN 1367-4803

Al-Shahrour, F.; Minguez, P.; Tarraga, J.; Medina, I.; Alloza, E.; Montaner, D. & Dopazo, J. (2007). FatiGO +: a functional profiling tool for genomic data. Integration of functional annotation, regulatory motifs and interaction data with microarray experiments. *Nucleic Acids Research,* Vol.35, pp. W91-96, ISSN 1362-4962

Alaiya, A. A.; Franzen, B.; Hagman, A.; Dysvik, B.; Roblick, U. J.; Becker, S.; Moberger, B.; Auer, G. & Linder, S. (2002). Molecular classification of borderline ovarian tumors using hierarchical cluster analysis of protein expression profiles. *International Journal of Cancer,* Vol.98, No.6, pp. 895-899, ISSN 0020-7136

Alfonso, P.; Nunez, A.; Madoz-Gurpide, J.; Lombardia, L.; Sanchez, L. & Casal, J. I. (2005). Proteomic expression analysis of colorectal cancer by two-dimensional differential gel electrophoresis. *Proteomics,* Vol.5, No.10, pp. 2602-2611, ISSN 1615-9853

Arighi, C. N.; Liu, H.; Natale, D. A.; Barker, W. C.; Drabkin, H.; Blake, J. A.; Smith, B. & Wu, C. H. (2009). TGF-beta signaling proteins and the Protein Ontology. *BMC Bioinformatics,* Vol.10, No. S5, pp. S3, ISSN 1471-2105

Arrell, D. K. & Terzic, A. Network systems biology for drug discovery. *Clinical Pharmacology and Therapeutics,* Vol.88, No.1, pp. 120-125, ISSN 1532-6535

Assenov, Y.; Ramirez, F.; Schelhorn, S. E.; Lengauer, T. & Albrecht, M. (2008). Computing topological parameters of biological networks. *Bioinformatics,* Vol.24, No.2, pp. 282-284, ISSN 1367-4811

Avila-Campillo, I.; Drew, K.; Lin, J.; Reiss, D. J. & Bonneau, R. (2007). BioNetBuilder: automatic integration of biological networks. *Bioinformatics*, Vol.23, No.3, pp. 392-393, ISSN 1367-4811

Bader, G. D. & Hogue, C. W. (2003). An automated method for finding molecular complexes in large protein interaction networks. *BMC Bioinformatics*, Vol.4, No., pp. 2, ISSN 1471-2105

Barabasi, A. L. & Oltvai, Z. N. (2004). Network biology: understanding the cell's functional organization. *Nat Rev Genet*, Vol.5, No.2, pp. 101-113, ISSN 1471-0056

Bhaskaran, N.; Lin, K. W.; Gautier, A.; Woksepp, H.; Hellman, U. & Souchelnytskyi, S. (2009). Comparative proteome profiling of MCF10A and 184A1 human breast epithelial cells emphasized involvement of CDK4 and cyclin D3 in cell proliferation. *Proteomics - Clinical Applications*, Vol.3, No.1, pp. 68-77, ISSN 1862-8354

Borrebaeck, C. A. & Wingren, C. (2007). High-throughput proteomics using antibody microarrays: an update. *Expert Review of Molecular Diagnostics*, Vol.7, No.5, pp. 673-686, ISSN 1744-8352

Boyle, E. I.; Weng, S.; Gollub, J.; Jin, H.; Botstein, D.; Cherry, J. M. & Sherlock, G. (2004). GO::TermFinder--open source software for accessing Gene Ontology information and finding significantly enriched Gene Ontology terms associated with a list of genes. *Bioinformatics*, Vol.20, No.18, pp. 3710-3715, ISSN 1367-4803

Breitkreutz, B. J.; Stark, C. & Tyers, M. (2002). Osprey: a network visualization system. *Genome Biology*, Vol.3, No.12, pp. PREPRINT0012, ISSN 1465-6914

Brennan, D. J.; O'Connor, D. P.; Rexhepaj, E.; Ponten, F. & Gallagher, W. M (2010). Antibody-based proteomics: fast-tracking molecular diagnostics in oncology. *Nature Review Cancer*, Vol.10, No.9, pp. 605-617, ISSN 1474-1768

Bussey, K. J.; Kane, D.; Sunshine, M.; Narasimhan, S.; Nishizuka, S.; Reinhold, W. C.; Zeeberg, B.; Ajay, W. & Weinstein, J. N. (2003). MatchMiner: a tool for batch navigation among gene and gene product identifiers. *Genome Biology*, Vol.4, No.4, pp. R27, ISSN 1465-6914

Caraux, G. & Pinloche, S. (2005). PermutMatrix: a graphical environment to arrange gene expression profiles in optimal linear order. *Bioinformatics*, Vol.21, No.7, pp. 1280-1281, ISSN 1367-4803

Carbon, S.; Ireland, A.; Mungall, C. J.; Shu, S.; Marshall, B. & Lewis, S. (2009). AmiGO: online access to ontology and annotation data. *Bioinformatics*, Vol.25, No.2, pp. 288-289, ISSN 1367-4811

Cerami, E. G.; Bader, G. D.; Gross, B. E. & Sander, C. (2006). cPath: open source software for collecting, storing, and querying biological pathways. *BMC Bioinformatics*, Vol.7, pp. 497, 1471-2105

Chen, G.; Gharib, T. G.; Wang, H.; Huang, C. C.; Kuick, R.; Thomas, D. G.; Shedden, K. A.; Misek, D. E.; Taylor, J. M.; Giordano, T. J.; Kardia, S. L.; Iannettoni, M. D.; Yee, J.; Hogg, P. J.; Orringer, M. B.; Hanash, S. M. & Beer, D. G. (2003). Protein profiles associated with survival in lung adenocarcinoma. *The Proceedings of National Academy of Sciences USA*, Vol.100, No.23, pp. 13537-13542, ISSN 0027-8424

Corona, G.; De Lorenzo, E.; Elia, C.; Simula, M. P.; Avellini, C.; Baccarani, U.; Lupo, F.; Tiribelli, C.; Colombatti, A. & Toffoli, G (2010). Differential proteomic analysis of hepatocellular carcinoma. *International Journal of Oncology,* Vol.36, No.1, pp. 93-99, ISSN 1791-2423

Cravatt, B. F.; Simon, G. M. & Yates, J. R., 3rd. (2007). The biological impact of mass-spectrometry-based proteomics. *Nature,* Vol.450, No.7172, pp. 991-1000, ISSN 1476-4687

Detwiler, L. T.; Mejino Jr, J. V.; Rosse, C. & Brinkley, J. F. (2003). Efficient web-based navigation of the Foundational Model of Anatomy. *AMIA Annual Symposium Proceedings,* pp. 829, ISSN 1942-597X

Dimmer, E. C.; Huntley, R. P.; Barrell, D. G.; Binns, D.; Draghici, S.; Camon, E. B.; Hubank, M.; Talmud, P. J.; Apweiler, R. & Lovering, R. C. (2008). The Gene Ontology - Providing a Functional Role in Proteomic Studies. *Proteomics,* Vol.8 No.23-24, ISSN 1615-9861

Draghici, S.; Khatri, P.; Bhavsar, P.; Shah, A.; Krawetz, S. A. & Tainsky, M. A. (2003). Onto-Tools, the toolkit of the modern biologist: Onto-Express, Onto-Compare, Onto-Design and Onto-Translate. *Nucleic Acids Research,* Vol.31, No.13, pp. 3775-3781, ISSN 1362-4962

Dudley, J. T. & Butte, A. J. (2009). Identification of discriminating biomarkers for human disease using integrative network biology. *Pacific Symposium of Biocomputing,* pp. 27-38, ISSN 1793-5091

Duncan, M. W. & Hunsucker, S. W. (2005). Proteomics as a tool for clinically relevant biomarker discovery and validation. *Experimental Biology and Medicine (Maywood),* Vol.230, No.11, pp. 808-817, ISSN 1535-3702

Eisen, M. B.; Spellman, P. T.; Brown, P. O. & Botstein, D. (1998). Cluster analysis and display of genome-wide expression patterns. *The Proceedings of National Academy of Sciences USA,* Vol.95, No.25, pp. 14863-14868, ISSN 0027-8424

Enright, A. J. & Ouzounis, C. A. (2001). BioLayout--an automatic graph layout algorithm for similarity visualization. *Bioinformatics,* Vol.17, No.9, pp. 853-854, ISSN 1367-4803

Funahashi, A.; Jouraku, A.; Matsuoka, Y. & Kitano, H. (2007). Integration of CellDesigner and SABIO-RK. *In Silico Biology,* Vol.7, No.S2, pp. S81-90, 1386-6338

Gao, J.; Ade, A. S.; Tarcea, V. G.; Weymouth, T. E.; Mirel, B. R.; Jagadish, H. V. & States, D. J. (2009). Integrating and annotating the interactome using the MiMI plugin for cytoscape. *Bioinformatics,* Vol.25, No.1, pp. 137-138, ISSN 1367-4811

Graves, P. R. & Haystead, T. A. (2002). Molecular biologist's guide to proteomics. *Microbiology and Molecular Biology Review,* Vol.66, No.1, pp. 39-63; ISSN 1092-2172

Gruber, T. R. (1993). A translation approach to portable ontologies. *Knowledge Acquisition,* Vol.5, pp. 199-220,

Hellman, K.; Alaiya, A. A.; Schedvins, K.; Steinberg, W.; Hellstrom, A. C. & Auer, G. (2004). Protein expression patterns in primary carcinoma of the vagina. *British Journal of Cancer,* Vol.91, No.2, pp. 319-326, ISSN 0007-0920

Khatri, P.; Draghici, S.; Ostermeier, G. C. & Krawetz, S. A. (2002). Profiling gene expression using onto-express. *Genomics,* Vol.79, No.2, pp. 266-270, ISSN 0888-7543

Kitano, H. (2002). Computational systems biology. *Nature,* Vol.420, No.6912, pp. 206-210, ISSN 0028-0836

Kitano, H. (2002). Systems biology: a brief overview. *Science*, Vol.295, No.5560, pp. 1662-1664, ISSN 1095-9203

Kohl, M.; Wiese, S. & Warscheid, B. (2011). Cytoscape: software for visualization and analysis of biological networks. *Methods in Molecular Biology*, Vol.696, pp. 291-303, ISSN 1940-6029

Kolker, E.; Higdon, R. & Hogan, J. M. (2006). Protein identification and expression analysis using mass spectrometry. *Trends in Microbiology*, Vol.14, No.5, pp. 229-235, 0966-842X

Kwoh, C. K. & Ng, P. Y. (2007). Network analysis approach for biology. *Cellular and Molecular Life Sciences*, Vol.64, No.14, pp. 1739-1751, ISSN 1420-682X

Li, L. S.; Kim, H.; Rhee, H.; Kim, S. H.; Shin, D. H.; Chung, K. Y.; Park, K. S.; Paik, Y. K. & Chang, J. (2004). Proteomic analysis distinguishes basaloid carcinoma as a distinct subtype of nonsmall cell lung carcinoma. *Proteomics*, Vol.4, No.11, pp. 3394-3400, ISSN 1615-9853

Lim, M. S. & Elenitoba-Johnson, K. S. (2004). Proteomics in pathology research. *Laboratory Investigation*, Vol.84, No.10, pp. 1227-1244, ISSN 0023-6837

Meunier, B.; Dumas, E.; Piec, I.; Bechet, D.; Hebraud, M. & Hocquette, J. F. (2007). Assessment of hierarchical clustering methodologies for proteomic data mining. *Journal of Proteome Research*, Vol.6, No.1, pp. 358-366, 1535-3893

Milli, A.; Cecconi, D.; Campostrini, N.; Timperio, A. M.; Zolla, L.; Righetti, S. C.; Zunino, F.; Perego, P.; Benedetti, V.; Gatti, L.; Odreman, F.; Vindigni, A. & Righetti, P. G. (2008). A proteomic approach for evaluating the cell response to a novel histone deacetylase inhibitor in colon cancer cells. *Biochimica et Biophysica Acta*, Vol.1784, No.11, pp. 1702-1710, ISSN 0006-3002

Morrissey, E. R. & Diaz-Uriarte, R. (2009). Pomelo II: finding differentially expressed genes. *Nucleic Acids Research*, Vol.37, No.Web Server issue, pp. W581-586, ISSN 1362-4962

Natale, D. A.; Arighi, C. N.; Barker, W. C.; Blake, J.; Chang, T. C.; Hu, Z.; Liu, H.; Smith, B. & Wu, C. H. (2007). Framework for a protein ontology. *BMC Bioinformatics*, Vol.8 Suppl 9, No., pp. S1, ISSN 1471-2105

Natale, D. A.; Arighi, C. N.; Barker, W. C.; Blake, J. A.; Bult, C. J.; Caudy, M.; Drabkin, H. J.; D'Eustachio, P.; Evsikov, A. V.; Huang, H.; Nchoutmboube, J.; Roberts, N. V.; Smith, B.; Zhang, J. & Wu, C. H. (2011). The Protein Ontology: a structured representation of protein forms and complexes. *Nucleic Acids Research*, Vol.39, pp. D539-545, ISSN 1362-4962

Perroud, B.; Lee, J.; Valkova, N.; Dhirapong, A.; Lin, P. Y.; Fiehn, O.; Kultz, D. & Weiss, R. H. (2006). Pathway analysis of kidney cancer using proteomics and metabolic profiling. *Molecular Cancer*, Vol.5, pp. 64, ISSN 1476-4598

Reeves, G. A.; Eilbeck, K.; Magrane, M.; O'Donovan, C.; Montecchi-Palazzi, L.; Harris, M. A.; Orchard, S.; Jimenez, R. C.; Prlic, A.; Hubbard, T. J.; Hermjakob, H. & Thornton, J. M. (2008). The Protein Feature Ontology: a tool for the unification of protein feature annotations. *Bioinformatics*, Vol.24, No.23, pp. 2767-2772, ISSN 1367-4811

Ringrose, J. H.; Jeeninga, R. E.; Berkhout, B. & Speijer, D. (2008). Proteomic studies reveal coordinated changes in T-cell expression patterns upon infection with human immunodeficiency virus type 1. *Journal of Virology,* Vol.82, No.9, pp. 4320-4330, ISSN 1098-5514

Scardoni, G.; Petterlini, M. & Laudanna, C. (2009). Analyzing biological network parameters with CentiScaPe. *Bioinformatics,* Vol.25, No.21, pp. 2857-2859, ISSN 1367-4811

Shannon, P.; Markiel, A.; Ozier, O.; Baliga, N. S.; Wang, J. T.; Ramage, D.; Amin, N.; Schwikowski, B. & Ideker, T. (2003). Cytoscape: a software environment for integrated models of biomolecular interaction networks. *Genome Research,* Vol.13, No.11, pp. 2498-2504, ISSN 1088-9051

Smith, B.; Ashburner, M.; Rosse, C.; Bard, J.; Bug, W.; Ceusters, W.; Goldberg, L. J.; Eilbeck, K.; Ireland, A.; Mungall, C. J.; Leontis, N.; Rocca-Serra, P.; Ruttenberg, A.; Sansone, S. A.; Scheuermann, R. H.; Shah, N.; Whetzel, P. L. & Lewis, S. (2007). The OBO Foundry: coordinated evolution of ontologies to support biomedical data integration. *Nature Biotechnology,* Vol.25, No.11, pp. 1251-1255, ISSN 1087-0156

Smoot, M. E.; Ono, K.; Ruscheinski, J.; Wang, P. L. & Ideker, T. (2011). Cytoscape 2.8: new features for data integration and network visualization. *Bioinformatics,* Vol.27, No.3, pp. 431-432, ISSN 1367-4811

Souchelnytskyi, S. (2005). Bridging proteomics and systems biology: what are the roads to be traveled? *Proteomics,* Vol.5, No.16, pp. 4123-4137, ISSN 1615-9853

Sturn, A.; Quackenbush, J. & Trajanoski, Z. (2002). Genesis: cluster analysis of microarray data. *Bioinformatics,* Vol.18, No.1, pp. 207-208, ISSN 1367-4803

Suehara, Y.; Kondo, T.; Fujii, K.; Hasegawa, T.; Kawai, A.; Seki, K.; Beppu, Y.; Nishimura, T.; Kurosawa, H. & Hirohashi, S. (2006). Proteomic signatures corresponding to histological classification and grading of soft-tissue sarcomas. *Proteomics,* Vol.6, No.15, pp. 4402-4409, ISSN 1615-9853

Taylor, C. F.; Paton, N. W.; Garwood, K. L.; Kirby, P. D.; Stead, D. A.; Yin, Z.; Deutsch, E. W.; Selway, L.; Walker, J.; Riba-Garcia, I.; Mohammed, S.; Deery, M. J.; Howard, J. A.; Dunkley, T.; Aebersold, R.; Kell, D. B.; Lilley, K. S.; Roepstorff, P.; Yates, J. R., 3rd; Brass, A.; Brown, A. J.; Cash, P.; Gaskell, S. J.; Hubbard, S. J. & Oliver, S. G. (2003). A systematic approach to modeling, capturing, and disseminating proteomics experimental data. *Nature Biotechnology,* Vol.21, No.3, pp. 247-254, ISSN 1087-0156

Thomas, S. & Bonchev, D. (2010). A survey of current software for network analysis in molecular biology. *Hum Genomics,* Vol.4, No.5, pp. 353-360, ISSN 1479-7364

Turner, K. E.; Kumar, H. R.; Hoelz, D. J.; Zhong, X.; Rescorla, F. J.; Hickey, R. J.; Malkas, L. H. & Sandoval, J. A. (2009). Proteomic analysis of neuroblastoma microenvironment: effect of the host-tumor interaction on disease progression. *Journal of Surgical Research,* Vol.156, No.1, pp. 116-122, ISSN 1095-8673

Wingren, C. & Borrebaeck, C. A. (2004). High-throughput proteomics using antibody microarrays. *Expert Rev Proteomics,* Vol.1, No.3, pp. 355-364, ISSN 1744-8387

Zhang, B.; Schmoyer, D.; Kirov, S. & Snoddy, J. (2004). GOTree Machine (GOTM): a web-based platform for interpreting sets of interesting genes using Gene Ontology hierarchies. *BMC Bioinformatics,* Vol.5, No., pp. 16, ISSN 1471-2105

Zhong, S.; Storch, K. F.; Lipan, O.; Kao, M. C.; Weitz, C. J. & Wong, W. H. (2004). GoSurfer: a graphical interactive tool for comparative analysis of large gene sets in Gene Ontology space. *Applied Bioinformatics,* Vol.3, No.4, pp. 261-264, ISSN 1175-5636

# nwCompare and AutoCompare Softwares for Proteomics and Transcriptomics Data Mining – Application to the Exploration of Gene Expression Profiles of Aggressive Lymphomas

Fréderic Pont[1,2,3], Marie Tosolini[1,2,3],
Bernard Ycart[4] and Jean-Jacques Fournié[1,2,3]
[1]INSERM UMR1037-Cancer Research Center of Toulouse
[2]ERL 5294 CNRS, BP3028, CHU Purpan, Toulouse,
[3]Université Toulouse III Paul-Sabatier, Toulouse,
[4]Laboratoire Jean Kuntzmann, CNRS UMR 5224,
Université Joseph Fourier, Grenoble
France

## 1. Introduction

The global protein and gene expression profiling technologies have revolutionized the study of normal and malignant cells. Transcriptomes permitted to delineate subtypes of B-cell lymphomas which were otherwise histologically and clinically undistinguishable. Although the data mining of proteomes or transcriptomes from these malignant cells can unveil new aspects of their biology, tools to simultaneously compare several samples are scarce. Here we depict nwCompare and Autocompare, two new freewares we developed with this aim, and examplify their use for the comparative data mining of transcriptomes from normal human B cells and B cell lymphomas such as follicular lymphomas (FL) and diffuse large B-cell lymphomas (DLBCL).

## 2. nwCompare software

Proteomics, transcriptomics and metabolomics implies the handling of a huge amount of data. Nano liquid chromatography combined with electrospray mass spectrometry enables the identification of hundreds of proteins in one complex sample whereas transcriptomics analyzes the expression level of about twenty thousand genes. It is very useful to be able to quickly compare lists of proteins, genes or molecules obtained from different patients, different pathological situations.

We designed nwCompare (Pont & Fournié, 2010), a software for n-way comparison of text files. nwCompare performs a line by line comparison of characters, thus, it can be quite useful to compare proteins names, gene names, molecules names, biological pathways names etc.

nwCompare has proven efficacy in proteomics to compare pathological situations (Pont & Fournié, 2010) or large-scale protein analysis (Pottiez & al., 2010).

The first versions of nwCompare were limited to analyse a maximum of 300 files, but, starting from version 3.20, this software is now only limited by the amount of memory of the computer. Moreover, a new feature has been introduced recently, by allowing the computation of a repartition table. It is thus possible to classify each file entry depending of its occurrence. nwCompare is light, very easy to use and enables users to run very complex comparisons just by selecting radio buttons, without learning any comparison syntax (Fig 1). nwCompare is a free software that can be download at:
http://www.ifr150.toulouse.inserm.fr/en/article.asp?id=264

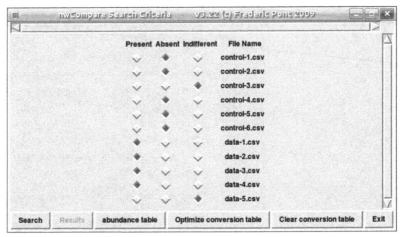

Fig. 1. Screenshot of nwCompare version 3.22 with the simultaneous comparison of eleven protein lists. This example shows the computation of the proteins present in four samples and absent in five controls with two files indifferents. The list of proteins matching those criteria is typically obtained in one second.

## 3. AutoCompare software

Autocompare freeware was developed as an evolution of nwCompare program to understand the biological significance of large lists of genes or proteins. This software takes as input any data text file and performs string comparisons by line of this file, with a collection of reference files (Fig 2). Then, for each of them, it computes the p-value of the comparison test from hypergeometric distribution tails, then corrects the raw p-values to account for multiple testing, using Bonferroni and Benjamini-Yekutieli methods (Fig 3). We provide AutoCompare with a starting collection of about 5000 genes reference lists based on GSEA (http://www.broadinstitute.org/gsea/) version 3.0 pathways and 162 protein lists based on PANTHER pathways (http://www.pantherdb.org/pathway/) (Mi & al., 2005). Indeed users can also implement in a very straightforward fashion any additional reference list (as .txt format) of their choice.

Autocompare was developed using the Perl programming language (Perl v5.10.1, http://www.perl.org/) and the R statistical programming language under the Linux operating system (ubuntu 10.04, http://www.ubuntu.com/). Autocompare is available for Linux and Windows (http://www.ifr150.toulouse.inserm.fr/en/article.asp?id=264), and runs on any operating system with Perl, either as a command line tool or with a graphical interface.

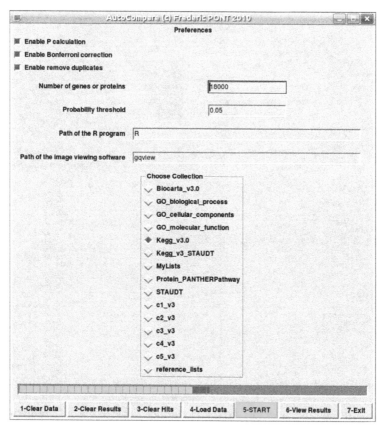

Fig. 2. Screenshot of AutoCompare version 2.31. The software is very easy to use and
provides users with biological significance of a large list of genes or proteins.

The main advantages of AutoCompare are that it is very easy to use, rapid and fully
automated, it works off line, the number of data files and reference files is only limited by
the available disk space, it is very easy to add personalized reference files. In addition, the
memory consumption of AutoCompare is very low because only two files are analyzed
simultaneously. As nwCompare, AutoCompare performs text file comparisons, so, any kind
of data files can potentially be analyzed with it, provided that reference files of the same
kind are used.

### 3.1 AutoCompare false discovery rate (FDR)

To calculate the FDR of AutoCompare, we randomized the genes comprised in the 186
genes lists from the Kegg library on the one hand and the genes comprised in the 3272 gene
lists from the Broad institute's GSEA C2' curated library. We then compared the
AutoCompare results obtained by querying the same experimental genes list with both the
randomized and the correct libraries. With the C2 library, the first false positive was
associated with a probability 44 times higher than the Bonferroni threshold. With the Kegg
library, the first false positive was associated with a probability 212 times higher than the

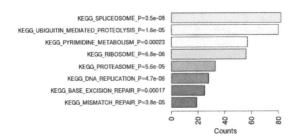

```
| Kegg_v3.8                            | Score | P value  | Counts / Total  (%) | Bonferroni | Benjamini/Yekutieli |
+--------------------------------------+-------+----------+---------------------+------------+---------------------+
| KEGG_SPLICEOSOME                     |  7    | 3.5e-08  | 82/128    ( 64 %)   | 6.51e-06   | 3.78e-05            |
| KEGG_DNA_REPLICATION                 |  5    | 4.7e-06  | 28/36     ( 78 %)   | 8.74e-04   | 2.45e-03            |
| KEGG_RIBOSOME                        |  5    | 6.8e-06  | 56/88     ( 64 %)   | 1.26e-03   | 2.45e-03            |
| KEGG_UBIQUITIN_MEDIATED_PROTEOLYSIS  |  5    | 1.6e-05  | 80/138    ( 58 %)   | 2.98e-03   | 4.32e-03            |
| KEGG_MISMATCH_REPAIR                 |  4    | 3.8e-05  | 19/23     ( 83 %)   | 7.07e-03   | 8.21e-03            |
| KEGG_PROTEASOME                      |  4    | 5.6e-05  | 33/48     ( 69 %)   | 1.04e-02   | 1.01e-02            |
| KEGG_BASE_EXCISION_REPAIR            |  4    | 0.00017  | 25/35     ( 71 %)   | 3.16e-02   | 2.62e-02            |
| KEGG_PYRIMIDINE_METABOLISM           |  4    | 0.00023  | 57/98     ( 58 %)   | 4.28e-02   | 3.10e-02            |
| KEGG_RNA_DEGRADATION                 |  3    | 0.00037  | 37/59     ( 63 %)   | 6.88e-02   | 4.44e-02            |
| KEGG_RNA_POLYMERASE                  |  3    | 0.00043  | 21/29     ( 72 %)   | 8.00e-02   | 4.64e-02            |
| KEGG_NUCLEOTIDE_EXCISION_REPAIR      |  3    | 0.00048  | 29/44     ( 66 %)   | 8.93e-02   | 4.71e-02            |
| KEGG_CELL_CYCLE                      |  3    | 0.0019   | 68/128    ( 53 %)   | 3.53e-01   | 1.71e-01            |
| KEGG_AMINOACYL_TRNA_BIOSYNTHESIS     |  3    | 0.0022   | 26/41     ( 63 %)   | 4.09e-01   | 1.83e-01            |
| KEGG_CHRONIC_MYELOID_LEUKEMIA        |  2    | 0.004    | 41/73     ( 56 %)   | 7.44e-01   | 3.09e-01            |
| KEGG_APOPTOSIS                       |  2    | 0.0043   | 48/88     ( 55 %)   | 8.00e-01   | 3.10e-01            |
| KEGG_PURINE_METABOLISM               |  2    | 0.0057   | 80/159    ( 50 %)   | 1.00e+00   | 3.85e-01            |
| KEGG_COLORECTAL_CANCER               |  2    | 0.013    | 34/62     ( 55 %)   | 1.00e+00   | 8.26e-01            |
| KEGG_PANCREATIC_CANCER               |  2    | 0.021    | 37/78     ( 53 %)   | 1.00e+00   | 1.00e+00            |
| KEGG_LYSOSOME                        |  2    | 0.021    | 60/121    ( 50 %)   | 1.00e+00   | 1.00e+00            |
| KEGG_HOMOLOGOUS_RECOMBINATION        |  2    | 0.022    | 17/28     ( 61 %)   | 1.00e+00   | 1.00e+00            |
| KEGG_ENDOMETRIAL_CANCER              |  2    | 0.031    | 28/52     ( 54 %)   | 1.00e+00   | 1.00e+00            |
| KEGG_N_GLYCAN_BIOSYNTHESIS           |  1    | 0.035    | 25/46     ( 54 %)   | 1.00e+00   | 1.00e+00            |
| KEGG_ENDOCYTOSIS                     |  1    | 0.047    | 85/183    ( 46 %)   | 1.00e+00   | 1.00e+00            |
| KEGG_PROTEIN_EXPORT                  |  1    | 0.055    | 14/24     ( 58 %)   | 1.00e+00   | 1.00e+00            |
| KEGG_PROSTATE_CANCER                 |  1    | 0.071    | 43/89     ( 48 %)   | 1.00e+00   | 1.00e+00            |
| KEGG_P53_SIGNALING_PATHWAY           |  1    | 0.077    | 34/69     ( 49 %)   | 1.00e+00   | 1.00e+00            |
```

Fig. 3. Example of results obtained with AutoCompare version 2.31. Top : histogram of the significant biological functions of a gene list data file. Counts indicate the number of genes found in the corresponding reference gene sets. Bottom: Table of the biological functions identified in a data file, sorted by statistical significance.

Bonferroni threshold. We thus implemented the FDR method of Benjamini and Yekutieli (Benjamini, Y. & Yekutieli, 2005, 2001) to adjust the P values in AutoCompare. This method controls the false discovery rate, the expected proportion of false discoveries among the rejected hypotheses. With the C2 library, the first false positive was associated with a probability 2.3 times higher than the probability of the first false positive. With the Kegg library by contrast, the first false positive was associated with a 119 times higher probability than for the first false positive. Hence, AutoCompare hits that are above the Bonferroni threshold are highly significant and without any false positive result. Furthermore, a good estimate of the FDR is thus given by the correction of Benjamini and Yekutieli.

## 4. Application of nwCompare and AutoCompare to explore the functional significance of gene expression profiles from normal B-cell subsets and of aggressive lymphomas

Normal differentiation of mature B lymphocytes comprises successive stages of maturation, in which naïve B cells reach germinal centers (GC) in lymph nodes and are activated by antigen to form centroblasts. These highly dividing GC centroblasts may further differentiate into centrocytes which, in turn, mature into either quiescent memory B cells or Ig-secreting plasmablasts which leave lymph nodes to home in bone arrow. Hence the normal B cell maturation in lymph nodes comprises the following sequence: Naïve > GC centroblasts > GC centrocytes > Memory cells, so we searched for the functions associated with the corresponding switches of gene expression signatures.

The transcriptome datasets (Affymetrix CEL files) GSE12195 (Compagno et al., 2009) and GSE15271 (Caron et al., 2009) produced with HG U133-Plus 2.0 platform were downloaded from the NCBI repository GEO database. Together, these comprised 27 normal B cell samples, including 4 naïve B cells, 9 tonsillar germinal center-derived centroblastic cells, 9 tonsillar germinal center-derived centrocytic and 5 memory B cells, 5 lymphoblastoid B-cell lines (B-LCL), 39 follicular lymphomas (FL) and 73 diffuse large B-cell lymphomas (DLBCL) (Caron et al., 2009; Compagno et al., 2009). The raw data from these 144 samples were log (base 2) transformed, normalized in batch by the RMA software and the 54676 probe sets were then reduced to a total of 20606 genes (HUGO symbols) by using the GSEA collapse function set on maximal probe mode (GSEA, http://www.broadinstitute.org/gsea), 18236 of which were fully annotated and thus kept for further study. The genes differentially expressed between two sample groups were defined using two-way Student's tests and $P<0.05$. These gene lists with one gene name per line were converted to text files and then uploaded in Autocompare. More than 4609 genes reference lists based on GSEA (http://www.broadinstitute.org/gsea/) version 3.0 pathways and 162 protein lists based on PANTHER pathways (http://www.pantherdb.org/pathway/) were collected. The differentially-expressed gene subsets were analyzed for enrichment in functionally-related genes among lists downloaded from the gene sets collection. Selective enrichment analysis was then computed with Autocompare using one-sided hypergeometric comparison tests, and False Discovery Rate corrections.

By using this approach, the genes that appeared differentially expressed ($P$-value <0.05) between respectively, naïve-and GC centroblast, GC centroblasts and GC centrocytes and between GC centrocytes and memory B cells were thus analyzed for functional significance by Autocompare using the KEGG library (V 3.0) of functional genesets in $H.$ $sapiens$. In this example, Autocompare performed the corresponding 1970 comparisons within 529 seconds. The GEP of the naïve-to-GC centroblast transition, the so-called "GC GEP signature", comprised 5516 differentially expressed genes. These latter witnessed of a significant increase of cell cycle ($P<10^{-20}$), DNA replication ($P<10^{-11}$), DNA damage and mismatch repair response, STAT3 signaling pathways together with reduced expression of genes for Krebs cycle metabolism and of IRF4-dependent plasmacytic differentiation genes (all with $P<10^{-5}$). Overall, this pattern reflected the unique differentiation program of B cells at the germinal center stage: a strong proliferation and high mutational activities which are both controlled by the Bcl-6 repressor, a program necessary for the clonal expansion of B-cells expressing mutated Ig. The profile of the centroblast-to-centrocyte GC transition comprised fewer differences (1966 differentially expressed genes) which corresponded to up-regulation of genes normally repressed by Bcl-6 ($P<10^{-8}$), hence reflecting the progressive disappearance of this transcriptional repressor. Finally, the GC centrocyte-to-memory B cell transition (5602 differentially expressed genes) showed a significant up-regulation of genes usually repressed by BLIMP-1, together with down-regulation of both cell cycle ($P<10^{-15}$), DNA replication ($P<10^{-11}$), DNA damage and mismatch repair response. This maturation profile, almost reverse to that of the N-to-GC centroblastic transition genes, indicated not only termination of the Bcl-6 dependent GC reaction but also a switch-off of the Blimp-1-dependent plasmacytic differentiation which, together, characterize quiescent memory cells. Hence the main physiological significance emerging from these comparisons is a signature of the germinal center reaction occurring in centroblasts: a unique combination of rapid proliferation and DNA remodeling (somatic hypermutations) without cell death.

## 4.1 Significance of gene signatures of non-Hodgkin's B cell lymphoma

Most non-Hodgkin's B cell lymphomas emerge from B cells in the germinal center (GC) stage however, by juxtaposing on their normal development the additional programs triggered by their genetic alterations. Accordingly, follicular lymphoma and diffuse large B-cell lymphoma are known to arise in normal GC B cells through genome alterations and mutations targeting genes controlling apoptotic cell death (BCL2, NFKB), differentiation (BCL6, MYC, BLIMP1, IRF4, CREBBP, EP300) or proliferation (BCR, CARD11, MYD88, NFKB, A20, STAT3) (for review, see Lenz & Staudt, 2010). The spectrum of oncogenes over-expressed and tumor suppressor genes under-expressed in each lymphoma translates to a corresponding profile which now defines the clinical subtype and contributes to predict outcome. We asked which of the 457 known human cancer genes (downloaded from the human cancer gene census (http:// www.sanger.ac.uk/ genetics/CGP/Census) were significantly deregulated in terms of either over-expressed oncogenes or down-regulated tumor suppressor genes in each lymphoma sample. Autocompare yielded such a list for each sample, and we then asked which were present in only one, in just two, in several, or in all of the these samples. Using corresponding requests, the 112 individual cancer gene lists were thus compared using nwCompare. This approach revealed that a total of 221 cancer genes were significantly deregulated in FL and DLBCL, among which 23 oncogenes were consistently upregulated in most (>75%) of the samples, like MAFB, ETV6 and COL1A1 which are strongly up-regulated in all (100%) of the samples (Figure 4). On the other hand 49 cancer genes were deregulated in only one or two patients, indicating these cancer genes are probably not driver cancer genes in the B-cell lymphomagenesis.

We then determined the complete set of genes which were differentially expressed by each individual lymphoma relative to the normal GC B cells ($P<0.05$). On average, 6735 genes were differentially expressed by each follicular lymphoma, 601 of which were shared by all FL. Although these comprised the hallmark over-expression (on average 20-fold) of the anti-apoptotic BCL2 gene, this 601 FL gene set also comprised other deregulated pathways. Using Autocompare, we found that these FL-deregulated pathways were significantly enriched for cytokine-cytokine receptor interactions (37/267 genes, P= 6.7e-12), complement and coagulation cascades (18/69 genes, P=5.1e-11), chemokine signalling (23/190 genes, P= 7.9e-07), ECM receptor interactions (14/84 genes, $P$=2.7e-06), focal adhesion (22/201 genes, P=7.2e-06), cell adhesion molecules (15/134 genes, $P$=0.0001), targets of BCL6 (7/19 genes, P= 3.5 e-06), targets of HIF-$\alpha$ (17/164 genes, P=0.0001). In addition, the FL GEP was significantly enriched in the previously depicted FL-type 1 (favourable outcome -associated) and type 2 (poor outcome-associated) immune response genes (respectively 10/40 genes, P=1.0e-06 and 6/23 genes, $P$=0.0001).

Within GC-type DLBCLs on average, 7365 genes were differentially expressed by each lymphoma, 376 of which were shared by all GC-type DLBCL. With ABC type DLBCL on average 7184 genes were differentially expressed by each ABC type DLBCL, 618 of which were shared by all ABC-type DLBCL.This suggested that DLBCL are more heterogeneous than FL, and that ABC-type DLBCL harbour the most genetically diversified profiles. The functional significance of both GC-type and ABC-type DLBCL gene sets comprised the same pathways as for FL plus the lysosome pathway (12/121 genes, P=5.1e-5).

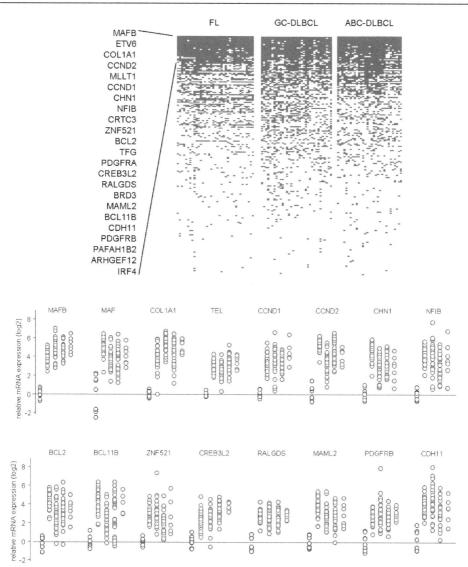

Top: Patterns of cancer genes differentially expressed (oncogenes over-expressed and tumor suppressor
genes under-expressed) in follicular lymphomas and diffuse large B cell lymphomas shows that
Follicular lymphomas are more homogeneous than DLBCL. Each column corresponds to a patient
sample and genes are lines. A blue dot means that the expression of the gene was deregulated for the
corresponding patient, a white dot means that the expression of the gene was similar to normal
individuals. MAFB, for example, is represented by an horizontal blue line, which mean that this gene
was deregulated in all patients. Bottom: Most significantly up-regulated oncogenes in aggressive
lymphomas. mRNA expression was normalized to the mean of normal samples (blue), compared to
patient's samples: follicular lymphomas (green), DLBCL (red circles), successively grouped as GC, ABC
and unclassified DLBCL subtypes, respectively.

Fig. 4. Pattern of oncogenes overexpressed in aggressive lymphomas.

## 5. Using AutoCompare with proteomics datasets

Proteomic scientists have two options to take benefit of AutoCompare with their proteomic datasets. The first option is straightforward: it is to use directly the starting collection of PANTHER protein pathways provided with AutoCompare. PANTHER protein pathways are built with Uniprot (http://www.uniprot.org/) protein accession numbers. If another protein database is to be used, the Protein Identifier Cross-Reference Service (PICR, http://www.ebi.ac.uk/Tools/picr/) can be applied to convert the data. Further, we present below two examples illustrating how AutoCompare can help data mining proteomes

Example 1: A virtual follicular lymphoma's proteome was created by converting genes up-regulated in follicular lymphoma (as depicted in §4) into protein accession numbers with PANTHER protein pathways. By using AutoCompare, this virtual proteome was then conveniently compared to a series of other proteomes, namely the whole PANTHER pathways proteome collection. Table 1 shows that the top ranking matches concerned proteins of apoptotic cell death, differentiation or proliferation (apoptosis, p53, p38 MAPK and Wnt pathways), focal adhesion (integrin and cadherin pathways), coagulation pathways, cytokines and chemokines signaling and immune response were differentially expressed in FL. In addition, angiogenesis (118/1231 proteins) and various growth factor signaling pathways (PDGF 65/938 proteins; VEGF 39/416 proteins; EGF 61/1071 proteins; IGF 32/276 proteins; FGF 53/978 proteins) were also enriched. Indeed in this example, these proteome comparisons matched with the results from transcriptome comparisons depicted in §4. Of note, the reverse strategy: converting protein accession numbers into gene names is also possible via the Protein Information Resource (PIR) (http://pir.georgetown.edu/pirwww/search/idmapping.shtml). Then, AutoCompare can be used with gene names, as described in § 4, taking advantage of the much larger collection of gene pathways provided with AutoCompare. The disadvantage of these conversion strategies however is that the original amount of data generally increases because of redundancy in databases and gene synonyms. Moreover, since most conversion tools do not filter results by taxonomy, this increase of non relevant data also augments the P values.

Example 2: Comparative analysis of experimental proteomes. The lymphoma cell line Karpas 299 was cultured in vitro for 48 hours in complete medium with and without the bisphosphonate drug zoledronate, the cells were isolated, their protein extract were prepared and the two resulting proteomes were analysed by mass spectrometry: briefly, the proteins were digested by trypsin, the peptides were analysed by nano-electrospray mass spectrometry and identified in SwissProt database using MASCOT (http://www.matrixscience.com/) software (unpublished results). AutoCompare allowed us to compare them to each other and to the proteomes listed the PANTHER pathways. This approach identified 52 matches between lymphoma proteins and one of the reference pathway proteomes. In control lymphoma cells for instance, Autocompare identified among others, 10 proteins of "cytoskeletal_regulation_by_Rho_GTPase" (O15144, O15145, O15511, P23528, P62736, P63261, P63267, P68032, P68133, Q5NBV3), 10 proteins involved in "inflammation mediated by chemokines and cytokines" and 6 proteins from the "Integrin_signalling_pathway". Of note, this approach also indicated that the 5 proteins P62736, P68032, P68133, Q13363 and Q969G3 expressed by the lymphoma cells in control conditions are involved in the Wnt pathway. By contrast, the proteome from cells treated with zoledronate only comprised the P68133 and Q13363 proteins from this pathway, suggesting the treatment had inhibited expression of the 3 others. Hence this example shows

how Autocompare can be used to pinpoint targeting of the morphogen Wnt cascade by the
bisphosphonate drug.

| Protein PANTHER Pathway | Counts/Total (%) |
| --- | --- |
| Integrin signalling pathway | 132/1175 (11%) |
| Inflammation mediated by chemokine and cytokine | 135/1417 (10%) |
| Angiogenesis | 118/1231 (10%) |
| Wnt signaling pathway | 142/2085 (7%) |
| Interleukin signaling pathway | 74/596 (12%) |
| Blood coagulation | 51/244 (21%) |
| Cadherin signaling pathway | 78/885 (9%) |
| Apoptosis signaling pathway | 74/839 (9%) |
| PDGF signaling pathway | 65/938 (7%) |
| p53 pathway | 49/548 (9%) |
| Toll receptor signaling pathway | 37/281 (13%) |
| Oxidative stress response | 39/349 (11%) |
| B-cell activation | 39/361 (11%) |
| TGF-beta signaling pathway | 56/810 (7%) |
| Heterotrimeric G-protein signaling pathway | 59/978 (6%) |
| VEGF signaling pathway | 39/416 (9%) |
| EGF receptor signaling pathway | 61/1071 (6%) |
| Heterotrimeric G-protein signaling pathway | 48/670 (7%) |
| Insulin IGF pathway-mitogen activated protein kinase | 32/276 (12%) |
| Interferon-gamma signaling pathway | 29/256 (11%) |
| Metabotropic glutamate receptor group II pathway | 36/423 (9%) |
| GABA-B receptor II signaling | 22/126 (17%) |
| T-cell activation | 43/631 (7%) |
| Cytoskeletal regulation by Rho GTPase | 44/674 (7%) |
| FGF signaling pathway | 53/978 (5%) |
| Enkephalin release | 29/280 (10%) |
| Endogenous cannabinoid signaling | 18/83 (22%) |
| Endothelin signaling pathway | 45/748 (6%) |
| Nicotinic acetylcholine receptor signaling pathway | 51/1011 (5%) |
| p38 MAPK pathway | 17/96 (18%) |

Table 1. Top rated PANTHER pathways identified by AutoCompare after conversion of
follicular lymphoma genes into protein accession numbers. Counts indicate the number of
genes found in the corresponding reference gene sets.

## 6. Conclusion

In conclusion, this example study shows how the use of Autocompare and nwCompare
enables users to get fast access to multidimensional comparisons and to the corresponding
analysis of large datasets such as proteomes and transcriptomes. We illustrated here this use
by the determination of oncogenes and functions involved in the biology of aggressive
human B cell lymphomas. Proteomics data sets (protein names, protein accession numbers)
can be compared directly in nwCompare since this software performs strings comparisons.

AutoCompare is provided with a starting collection of PANTHER protein pathways for a direct analysis of proteomic datasets. Proteomics users can additionally take advantage of AutoCompare large gene starting database of about 5500 pathways by converting protein names into gene names.

## 7. Acknowledgements

Work in JJF's lab is supported by institutional grants from INSERM, Université de Toulouse 3 and CNRS, as well as by grants from Institut National du Cancer (contracts RITUXOP and V9V2TER). We thank L. Pasqualucci (Columbia University, NY) for kindly providing us with clinical classifications of the lymphoma samples from GSE12195 dataset.

## 8. References

Benjamini, Y., & Yekutieli, D. (2005). Quantitative trait loci analysis using the false discovery rate. *Genetics*, 171, pp 783-790, Print ISSN: 0016-6731; Online ISSN: 1943-2631

Benjamini, Y., & Yekutieli, D. (2001). The control of the false discovery rate in multiple testing under dependency. *Ann. Stat.*, 29, 4, pp 1165-1188, ISSN 0090-5364

Caron, G., Le Gallou, S., Lamy, T., Tarte, K. & Fest, T. (2009). CXCR4 expression functionally discriminates centroblasts versus centrocytes within human germinal center B cells. *J Immunol*. 182, pp 7595-7602.

Compagno, M., Lim, W. K., Grunn, A., Nandula, S. V., Brahmachary, M., Shen, Q., Bertoni, F., Ponzoni, M., Scandurra, M., Califano, A., Bhagat, G., Chadburn, A., Dalla-Favera, R. & Pasqualucci, L. (2009). Mutations of multiple genes cause deregulation of NF-kappaB in diffuse large B-cell lymphoma. *Nature*, 459, pp 717-721.

Côté, R.G., Jones, P., Martens, L., Kerrien, S., Reisinger, F., Lin, Q., Leinonen, R., Apweiler, R. & Hermjakob, H. (2007). The Protein Identifier Cross-Referencing (PICR) service: reconciling protein identifiers across multiple source databases. *BMC Bioinformatics*, 8, pp 401-414.

Lenz, G. and Staudt, L. (2010). Aggressive lymphomas. *The New England Journal of Medicine,*. 362, pp 1419-1429.

Mi, H., Lazareva-Ulitsky, B., Loo, R., Kejariwal, A., Vandergriff, J., Rabkin, S., Guo, N., Muruganujan, A., Doremieux, O., Campbell, M. J., Kitano, H. & Thomas* P. D. (2005). The PANTHER database of protein families, subfamilies, functions and pathways. *Nucl. Acids Res*, 33, suppl 1, D284-D288.

Pont, F. & Fournié, JJ. (2010). Sorting protein lists with nwCompare: a simple and fast algorithm for n-way comparison of proteomic data files. *Proteomics*, 10, 5, March 2010, pp 1091-1094. ISSN: 1615-9861.

Pottiez, G., Deracinois, B., Duban-Deweer, S., Cecchelli, R., Fenart, L., Karamanos, Y. & Flahaut, C. (2010). A large-scale electrophoresis- and chromatography-based determination of gene expression profiles in bovine brain capillary endothelial cells after the re-induction of blood-brain barrier properties. *Proteome Sci.*, 15, 8, November 2010, pp 57. ISSN: 1477-5956.

# Permissions

The contributors of this book come from diverse backgrounds, making this book a truly international effort. This book will bring forth new frontiers with its revolutionizing research information and detailed analysis of the nascent developments around the world.

We would like to thank Hon-Chiu Eastwood Leung, Ph.D., for lending his expertise to make the book truly unique. He has played a crucial role in the development of this book. Without his invaluable contribution this book wouldn't have been possible. He has made vital efforts to compile up to date information on the varied aspects of this subject to make this book a valuable addition to the collection of many professionals and students.

This book was conceptualized with the vision of imparting up-to-date information and advanced data in this field. To ensure the same, a matchless editorial board was set up. Every individual on the board went through rigorous rounds of assessment to prove their worth. After which they invested a large part of their time researching and compiling the most relevant data for our readers. Conferences and sessions were held from time to time between the editorial board and the contributing authors to present the data in the most comprehensible form. The editorial team has worked tirelessly to provide valuable and valid information to help people across the globe.

Every chapter published in this book has been scrutinized by our experts. Their significance has been extensively debated. The topics covered herein carry significant findings which will fuel the growth of the discipline. They may even be implemented as practical applications or may be referred to as a beginning point for another development. Chapters in this book were first published by InTech; hereby published with permission under the Creative Commons Attribution License or equivalent.

The editorial board has been involved in producing this book since its inception. They have spent rigorous hours researching and exploring the diverse topics which have resulted in the successful publishing of this book. They have passed on their knowledge of decades through this book. To expedite this challenging task, the publisher supported the team at every step. A small team of assistant editors was also appointed to further simplify the editing procedure and attain best results for the readers.

Our editorial team has been hand-picked from every corner of the world. Their multi-ethnicity adds dynamic inputs to the discussions which result in innovative outcomes. These outcomes are then further discussed with the researchers and contributors who give their valuable feedback and opinion regarding the same. The feedback is then collaborated with the researches and they are edited in a comprehensive manner to aid the understanding of the subject.

Apart from the editorial board, the designing team has also invested a significant amount of their time in understanding the subject and creating the most relevant covers. They scrutinized every image to scout for the most suitable representation of the subject and create an appropriate cover for the book.

The publishing team has been involved in this book since its early stages. They were actively engaged in every process, be it collecting the data, connecting with the contributors or procuring relevant information. The team has been an ardent support to the editorial, designing and production team. Their endless efforts to recruit the best for this project, has resulted in the accomplishment of this book. They are a veteran in the field of academics and their pool of knowledge is as vast as their experience in printing. Their expertise and guidance has proved useful at every step. Their uncompromising quality standards have made this book an exceptional effort. Their encouragement from time to time has been an inspiration for everyone.

The publisher and the editorial board hope that this book will prove to be a valuable piece of knowledge for researchers, students, practitioners and scholars across the globe.

# List of Contributors

**Amanda Nouwens and Stephen Mahler**
The University of Queensland, Australia

**Fernanda Salvato**
Universidade de São Paulo, Escola Superior de Agricultura Luiz de Queiroz, Brazil

**Mayra Costa da Cruz Gallo de Carvalho**
Empresa Brasileira de Pesquisa Agropecuária (EMBRAPA), Brazil

**Aline de Lima Leite**
Universidade de São Paulo, Faculdade de Odontologia de Bauru, Brazil

**Hanne Kolsrud Hustoft, Helle Malerod, Steven Ray Wilson, Leon Reubsaet, Elsa Lundanes and Tyge Greibrokk**
University of Oslo, Norway

**Valentina Fiorilli and Raffaella Balestrini**
Istituto per la Protezione delle Piante del CNR and Dipartimento Biologia Vegetale dell'Università di Torino, Torino, Italy

**Vincent P. Klink**
Department of Biological Sciences, Harned Hall, Mississippi State University - Mississippi State, USA

**Hiroshi Yamaguchi**
Measurement Solution Research Center, National Institute of Advanced Industrial Science and Technology, Japan
Liberal Arts Education Center, Aso campus, Tokai University, Japan

**Masaya Miyazaki and Hideaki Maeda**
Measurement Solution Research Center, National Institute of Advanced Industrial Science and Technology, Japan
Interdisciplinary Graduate School of Engineering Science, Kyusyu University, Japan

**Karen A. Sap and Jeroen A. A. Demmers**
Erasmus University Medical Center, The Netherlands

**Igor Kučera**
Department of Biochemistry, Faculty of Science, Masaryk University, Brno, Czech Republic

**Pavel Bouchal**
Department of Biochemistry, Faculty of Science, Masaryk University, Brno, Czech Republic
Regional Centre for Applied Molecular Oncology, Masaryk Memorial Cancer Institute, Brno, Czech Republic

**Robert Stein**
I&B Informatics and Biology, Berlin, Germany

**Zbyněk Zdráhal**
Core Facility – Proteomics, Central European Institute of Technology, Masaryk University, Brno, Czech Republic

**Peter R. Jungblut**
Max Planck Institute for Infection Biology, Core Facility Protein Analysis, Berlin, Germany

**Chengjian Xie**
Chongqing Normal University, Chongqing, China

**Xiaowen Wang and Anping Sui**
Southwest University, Chongqing, China

**Xingyong Yang**
Chongqing Normal University, Chongqing, China
Southwest University, Chongqing, China

**Parminder Kaur and Mark R. Chance**
Center for Proteomics and Bioinformatics, Case Western Reserve University, Cleveland, OH
NeoProteomics, Inc., Cleveland, OH, USA

**Kah Wai Lin, Min Jia and Serhiy Souchelnytskyi**
Department of Oncology-Pathology, Karolinska Institutet, Stockholm, Sweden

**Fréderic Pont, Marie Tosolini and Jean-Jacques Fournié**
INSERM UMR1037-Cancer Research Center of Toulouse, France
ERL 5294 CNRS, BP3028, CHU Purpan, Toulouse, France
Université Toulouse III Paul-Sabatier, Toulouse, France

**Bernard Ycart**
Laboratoire Jean Kuntzmann, CNRS UMR 5224, Université Joseph Fourier, Grenoble, France

Printed in the USA
CPSIA information can be obtained
at www.ICGtesting.com
JSHW011410221024
72173JS00003B/485

9 781632 394415